New Challenges and Perspectives in Neurology and Autonomic Disorders

New Challenges and Perspectives in Neurology and Autonomic Disorders

Guest Editors

Svetlana Blitshteyn
Ilene Ruhoy
Jennifer Robblee

MDPI

Basel • Beijing • Wuhan • Barcelona • Belgrade • Novi Sad • Cluj • Manchester

Guest Editors
Svetlana Blitshteyn
Department of Neurology
University at Buffalo
Buffalo, NY
USA

Ilene Ruhoy
Mount Sinai South Nassau
Oceanside, NY
USA

Jennifer Robblee
Barrow Neurological Institute
Phoenix, AZ
USA

Editorial Office
MDPI AG
Grosspeteranlage 5
4052 Basel, Switzerland

This is a reprint of the Special Issue, published open access by the journal *Journal of Personalized Medicine* (ISSN 2075-4426), freely accessible at: https://www.mdpi.com/journal/jpm/special_issues/1IJRC0UDMI.

For citation purposes, cite each article independently as indicated on the article page online and as indicated below:

Lastname, A.A.; Lastname, B.B. Article Title. *Journal Name* **Year**, *Volume Number*, Page Range.

ISBN 978-3-7258-5293-2 (Hbk)
ISBN 978-3-7258-5294-9 (PDF)
https://doi.org/10.3390/books978-3-7258-5294-9

Cover image courtesy of Svetlana Blitshteyn

Contents

About the Editors

Svetlana Blitshteyn

Svetlana Blitshteyn, MD, FAAN, FANA, is the Director and Founder of the Dysautonomia Clinic and a Clinical Associate Professor of Neurology at the University at Buffalo Jacobs School of Medicine and Biomedical Sciences. She completed her neurology training at the Mayo Clinic Graduate School of Medicine. She currently serves on the NIH RECOVER-Treating Long COVID Neurological Agents committee. Previously, she was appointed as the Clinical Lead for the Autonomic Section Writing Group for the Multi-Disciplinary PASC/Long COVID Collaborative. She was also selected to contribute to the definition of Long COVID for the National Academies of Sciences, Engineering, and Medicine as part of a focus group. She is a Fellow of the American Academy of Neurology and the American Neurological Association. Dr. Blitshteyn co-authored a popular patient handbook called "*POTS—Together We Stand; Riding the Waves of Dysautonomia*" and has been interviewed and quoted by numerous media outlets, including the Washington Post, New York Times, US News and World Report, Newsweek, Guardian, New Scientist, Medscape, and others. She received the 2025 Research All-Star in Neurology recognition for her outstanding contributions to healthcare research, ranking in the top 3% nationwide; the Women of Distinction in Healthcare 2024 Award from the New York State Assembly; the 2022 Dysautonomia International Physician of the Year Award, and other awards. Her research focuses on POTS, Long COVID, Ehlers–Danlos syndrome, and women's health.

Ilene Ruhoy

Ilene Ruhoy, MD, PhD, is a board-certified neurologist and an environmental toxicologist who specializes in chronic and complex illness. She serves as the Medical Director of the Chiari EDS Center at Mount Sinai South. She graduated from the University of Pittsburgh School of Medicine and completed her residency in neurology at the University of Washington, where she also completed additional fellowship training in neuromuscular disorders. She earned a PhD in Environmental Toxicology at the University of Nevada, working directly with the Environmental Protection Agency (EPA) on her dissertation topic of "Pharmaceutical Residues in the Water." Dr. Ruhoy has also completed a fellowship in Integrative Medicine with Dr. Andrew Weil at the University of Arizona. Dr. Ruhoy is a co-editor of the Special Issue of Neurology and Connective Tissue for Frontiers in *Neurology*. She has also been a co-editor of *Integrative Neurology* published by Oxford Press and a co-editor of Preventive Neurology, of the *Seminars in Neurology* series. She is the Chair of the Neurology Working Committee for the Ehlers-Danlos Society and is also a part of the Neurology Working Group for the national apheresis organization, AFSA.

Jennifer Robblee

Jennifer Robblee, MD, is a neurologist and associate professor in the Department of Neurology at Barrow Neurological Institute. Dr. Robblee's expertise includes the diagnosis and treatment of migraine and other headache disorders. She is a member of the American Academy of Neurology, American Headache Society, the College of Physicians and Surgeons of Ontario, and the International Headache Society.

Journal of *Personalized Medicine*

Editorial

New Challenges and Perspectives in Neurology and Autonomic Disorders: A Leap Forward

Svetlana Blitshteyn [1,2,*] and Ilene Ruhoy [3]

1 Department of Neurology, School of Medicine and Biomedical Sciences, University of Buffalo Jacobs, Buffalo, NY 14203, USA
2 Dysautonomia Clinic, Williamsville, NY 14221, USA
3 Department of Neurology, Mount Sinai South Nassau, Oceanside, NY 11572, USA
* Correspondence: sb25@buffalo.edu

"Nothing in life is to be feared, it is only to be understood. Now is the time to understand more so that we may fear less."

—Marie Curie

The autonomic nervous system, which consists of the sympathetic, parasympathetic, and enteric divisions, is an integral part of the central and peripheral nervous systems and controls homeostasis, blood flow, and responses to internal and external stimuli. Disorders of the autonomic nervous system—both common, such as postural orthostatic tachycardia syndrome (POTS), neurocardiogenic syncope, and orthostatic hypotension (OH); and rare, such as multiple system atrophy, amyloid neuropathy, and familial dysautonomia—have been an evolving area of research in basic and translational science as well as in clinical practice. More recently, the COVID-19 pandemic further underscored the need to elucidate the neurologic and autonomic mechanisms of post-infectious syndromes. To this end, the new frontier in neurology and autonomic disorders, as well as the mechanistic interplay between a wide range of neurologic conditions and autonomic dysfunctions, present an exciting opportunity for groundbreaking discoveries.

In this Special Issue, we aimed to collect original research, reviews, hypothesis, perspectives, and opinions on autonomic disorders and how it affects various medical subspecialties, including neurology, cardiology, infection-associated chronic illnesses, headache medicine, psychiatry, and others. We were particularly interested in studies and reviews of the potential biomarkers and identification of effective diagnostic and therapeutic approaches in patients with complex neurologic and autonomic disorders and how they advance our understanding of these disorders, in addition to helping us expand our diagnostic and therapeutic capabilities in clinical practice.

It has been known from clinical observations and some experimental studies that the autonomic nervous system extends beyond the regulation of the target organs by the parasympathetic, sympathetic, and enteric nervous systems and that it closely communicates with the immunologic system and inflammatory pathways. David Goldstein, MD, Ph.D., Chief of Autonomic Section at the National Institutes of Health, provides a perspective on the "extended" autonomic system (EAS) and the "homeostat" theory as applied to the pathophysiology and potential treatments of dysautonomia (contribution 1). He emphasizes that the ANS may include neuroendocrine, immune/inflammatory, and central components and that comparators in the form of thermostat, glucostat, carbistat, and barostat exist that regulate different variables, such as core temperature, blood glucose, blood gases, and delivery of blood to the brain. He presents the homeostat theory and how it applies to EAS with specific examples of pediatric, adolescent/adult, and geriatric forms of dysautonomia and argues that computer modeling has the potential to lead to individualized treatments and outcomes (contribution 1).

The neuropsychiatric manifestations of systemic disease are an under-represented area of research in neurology that deserve increased research interest, education time, and neurology training. Weinstock et al. reported a case series of eight patients with mast cell activation syndrome (MCAS)—a multisystemic immunologic disorder with an estimated prevalence of 17%—who experienced significant neuropsychiatric disorders that were refractory to standard therapies. Five patients had depression, five had generalized anxiety disorder, and four had a panic disorder (contribution 2). All eight patients were subsequently diagnosed with MCAS; six out of eight patients had comorbid autonomic disorders with the most common being POTS, and four had hypermobile Ehlers–Danlos syndrome (h-EDS). All patients experienced significant improvement in their neuropsychiatric and multisystemic symptoms after mast-cell-directed therapy was implemented, which included antihistamines, mast-cell-stabilizing agents and a low-histamine diet (contribution 2). This case series illustrates the systemic nature of common neurologic and psychiatric disorders, which need to be identified and treated.

Migraine is one of the most comorbidities of POTS, with both disorders being heavily influenced by sex hormones. Godley III et al. review how sex hormones affect migraine with the help of interdisciplinary research scientists that focused on examining estrogen and oxytocin while noting that progesterone, testosterone, and vasopressin were less well-studied (contribution 3). They conclude that progress in research on the effects of hormones on the nervous system has been slow and that substantial gaps exist in our understanding of the complex roles sex hormones play in migraine (contribution 3). Increased funding, interdisciplinary research efforts, and exploring therapeutic agents, such as oxytocin delivered via nasal spray, could advance the science and therapeutic implications of sex hormones in women with migraine at various stages of life.

Long COVID-19 highlighted a wide gap in our understanding and clinical approach to patients with post-acute infection syndromes, which commonly include POTS and myalgic encephalomyelitis/chronic fatigue syndrome (ME/CFS). However, with the renewed interest and investment in the field of infection-associated chronic illnesses, we have made significant progress in our understanding of the complex pathophysiology of post-acute sequelae of SARS-CoV-2, which includes autonomic dysfunction, immune dysregulation, endothelial disturbance, vascular and neuropathic changes, microbiome alteration, microglial activation, blood–brain barrier disruption and other pathologic manifestations [1]. Importantly, Davenport et al. emphasize that ong COVID-19 is not a functional neurologic disorder (FND) (contribution 4). FND is previously known as conversion disorder and hysteria—a condition rooted in psychosocial etiology and distorted sense of agency and emotional processing, which is commonly treated with FND-targed psychotherapy and physical therapy. The authors assert that the vast majority of patients with long COVID-19 do not have FND, but do have dysautonomia and ME/CFS, which should not be mislabeled with FND as pathophysiology, diagnostic tests, physical exam, and treatment approaches are significantly different between these disorders (contribution 4).

Adding to the expanding science on long COVID-19, Tabacof et al. reviewed echocardiograms of over 200 patients with post-COVID-19 dysautonomia and queried if these symptoms may be cardiogenic (contribution 5). They found that most patients did not show evidence of cardiac abnormalities on echocardiography. Interestingly, they found that patients with post-COVID-19 dysautonomia had lower stroke volume than in an unclassified subgroup and that stroke volume and left-ventricular end-diastolic volume were smaller in those reporting decreased physical activity after COVID-19 (contribution 5). Similar findings were identified in patients with POTS before the COVID-19 pandemic [2,3].

Hypercoagulable state has been found in many patients with long COVID-19 and some patients with POTS. Kell and Pretorius et al. argue that fibrinaloid microclots may be important in the pathophysiology of POTS through their ability to block the flow of blood through microcapillaries and thus cause tissue hypoperfusion (contribution 6). Amyloids are known to be membrane disruptors and may affect the autonomic nerve fibers. Previously, they showed the presence of microclots in patients with long COVID-19 [4]—a finding that was recently confirmed by another study demonstrating that fibrin drives the thromboinflammation and neuropathology after COVID-19 infection [5]. It remains to be determined whether the same mechanisms are involved in POTS, especially in the context of post-acute infection syndromes.

Current research highlight autoimmunity as an important pathophysiologic mechanism of POTS and its numerous comorbidities, including gastrointestinal disorders. Nakane et al. examined patients diagnosed with functional gastrointestinal disorders and found that among 11 patients with irritable bowel syndrome and functional dyspepsia, 4 had anti-ganglionic nicotinic acetylcholine receptor antibodies measured via luciferase immunoprecipitation system assay, with 3 also having dry eyes and dry mouth, while there were no such symptoms in antibody-negative group (contribution 7). Further studies are needed to determine the prevalence of these and other antibodies in patients with functional gastrointestinal disorders.

Continuing the important topic of autoimmunity in autonomic disorders, Pena et al. offered a comprehensive literature review on a variety of autoantibodies and immunomodulatory therapies that have been described in patients with POTS and OH (contribution 8). They highlight the existing studies and case series that demonstrate the presence of antinuclear, anti-phospholipid, alpha and beta adrenergic, cholinergic, and angiotensin II type I autoantibodies associated with POTS and OH. Importantly, case reports and series suggest that immunotherapy with intravenous and subcutaneous immunoglobulin as well as plasmapheresis can be beneficial in patients with severe POTS refractory to standard therapies (contribution 8). Large clinical trials, including the NIH RECOVER-AUTONOMIC trial assessing the benefits of IVIG, are currently in progress to determine the efficacy of these therapies in patients with post-COVID POTS and autonomic dysfunction [6].

Diagnostic testing is an integral part of the clinical evaluation and diagnostic criteria of autonomic disorders. Jason et al. conducted a study of 193 patients with ME/CFS using a tilt table test whereas 32.5% of patients in this cohort tested positive for POTS or OH (contribution 9). The participants with either of these two common autonomic disorders were found to have more problems with sleep and post-exertional malaise as well as greater physical and health function limitations. These findings highlight the need for further understanding of the etiology of symptoms that have been ascribed to POTS or OH, how the symptoms may or may not interfere in the interpretation of the TTT, and what other tests (that are more sensitive and specific for autonomic dysfunction) should be developed for patients with clear autonomic dysfunction, but a negative tilt table test.

Perfecting simple and non-invasive means of testing for objective data and improved diagnosis and thus treatment options for patients with Parkinson's disease, Fernando et al. found that having patients with Parkinson's disease take their blood pressure at home twice daily in lying and standing positions over 5 days greatly improved identification of blood pressure disturbance, including OH, in comparison to a single in-office measurement (contribution 10). This may help with early recognition of OH and other blood pressure disturbances, as well as implementation of interventions to mitigate dysautonomia, in patients with Parkinson's disease—one of the most common movement disorders seen in neurology clinics.

This Special Issue is only a tiny particle in the universe of the unknown about the brain and the autonomic nervous system. Nevertheless, we hope it can serve as an important source of information that generates excitement, curiosity, and further research in neurology, neuroscience, and interdisciplinary specialties. It has been an honor and a privilege for us to serve as the Guest Editors of this Special Issue, and to read, edit, and learn from the innovative, cutting-edge contributions, written by some of the best researchers, clinicians, and world-renowned experts in the field. As neurology and autonomic disorders advance and as we continue to face the COVID-19 pandemic with increased prevalence of post-COVID-19 complications, investigating the pathophysiology, diagnostic tests, and therapeutic options for complex neurologic and autonomic disorders becomes our top priority. It is a leap forward that we must take in our quest for scientific inquiry and advancement in medicine, science, patient care, and public health.

Author Contributions: S.B. wrote the original draft; S.B. and I.R. revised and edited the manuscript. All authors have read and agreed to the published version of the manuscript.

Acknowledgments: We would like to thank all of the contributors and co-authors who submitted their papers to this Special Issue. We would like to also thank the patients, caregivers and study participants who selflessly provided us with their case histories, photos, and participation in the studies in order to contribute to science. Finally, we appreciate the work of the reviewers, editors, staff, and the publisher who offered their expertise and assistance in making this Special Issue a success.

Conflicts of Interest: The authors declare no conflicts of interest.

List of Contributions:

1. Goldstein, D.S. Linking the Extended Autonomic System with the Homeostat Theory: New Perspectives about Dysautonomias. *J. Pers. Med.* **2024**, *14*, 123. https://doi.org/10.3390/jpm14010123.
2. Weinstock, L.B.; Nelson, R.M.; Blitshteyn, S. Neuropsychiatric Manifestations of Mast Cell Activation Syndrome and Response to Mast-Cell-Directed Treatment: A Case Series. *J. Pers. Med.* **2023**, *13*, 1562. https://doi.org/10.3390/jpm13111562.
3. Godley, F., III; Meitzen, J.; Nahman-Averbuch, H.; O'Neal, M.A.; Yeomans, D.; Santoro, N.; Riggins, N.; Edvinsson, L. How Sex Hormones Affect Migraine: An Interdisciplinary Preclinical Research Panel Review. *J. Pers. Med.* **2024**, *14*, 184. https://doi.org/10.3390/jpm14020184.
4. Davenport, T.E.; Blitshteyn, S.; Clague-Baker, N.; Davies-Payne, D.; Treisman, G.J.; Tyson, S.F. Long COVID Is Not a Functional Neurologic Disorder. *J. Pers. Med.* **2024**, *14*, 799. https://doi.org/10.3390/jpm14080799.
5. Tabacof, L.; Wood, J.; Breyman, E.; Tosto-Mancuso, J.; Kelly, A.; Wilkey, K.; Zhang, C.; Putrino, D.; Kontorovich, A. Dysautonomia, but Not Cardiac Dysfunction, Is Common in a Cohort of Individuals with Long COVID. *J. Pers. Med.* **2023**, *13*, 1606. https://doi.org/10.3390/jpm13111606.
6. Kell, D.B.; Khan, M.A.; Kane, B.; Lip, G.Y.H.; Pretorius, E. Possible Role of Fibrinaloid Microclots in Postural Orthostatic Tachycardia Syndrome (POTS): Focus on Long COVID. *J. Pers. Med.* **2024**, *14*, 170. https://doi.org/10.3390/jpm14020170.
7. Nakane, S.; Mukaino, A.; Okumura, Y.; Hirosawa, H.; Higuchi, O.; Matsuo, H.; Kainuma, M.; Nakatsuji, Y. The Presence of Ganglionic Acetylcholine Receptor Antibodies in Sera from Patients with Functional Gastrointestinal Disorders: A Preliminary Study. *J. Pers. Med.* **2024**, *14*, 485. https://doi.org/10.3390/jpm14050485.
8. Pena, C.; Moustafa, A.; Mohamed, A.-R.; Grubb, B. Autoimmunity in Syndromes of Orthostatic Intolerance: An Updated Review. *J. Pers. Med.* **2024**, *14*, 435. https://doi.org/10.3390/jpm14040435.
9. Jason, L.A.; McGarrigle, W.J.; Vermeulen, R.C.W. The Head-Up Tilt Table Test as a Measure of Autonomic Functioning among Patients with Myalgic Encephalomyelitis/Chronic Fatigue Syndrome. *J. Pers. Med.* **2024**, *14*, 238. https://doi.org/10.3390/jpm14030238.
10. Fernando, C.V.; Osborn, S.; Horne, M. At-Home Blood Pressure Measurements Provide Better Assessments of Orthostatic Hypotension in Parkinson's Disease. *J. Pers. Med.* **2023**, *13*, 1324. https://doi.org/10.3390/jpm13091324.

References

1. Al-Aly, Z.; Davis, H.; McCorkell, L.; Soares, L.; Wulf-Hanson, S.; Iwasaki, A.; Topol, E.J. Long COVID science, research and policy. *Nat. Med.* **2024**, *30*, 2148–2164. [CrossRef]
2. Masuki, S.; Eisenach, J.H.; Schrage, W.G.; Johnson, C.P.; Dietz, N.M.; Wilkins, B.W.; Sandroni, P.; Low, P.A.; Joyner, M.J. Reduced stroke volume during exercise in postural tachycardia syndrome. *J. Appl. Physiol.* **2007**, *103*, 1128–1135. [CrossRef]
3. Oldham, W.M.; Lewis, G.D.; Opotowsky, A.R.; Waxman, A.B.; Systrom, D.M. Unexplained exertional dyspnea caused by low ventricular filling pressures: Results from clinical invasive cardiopulmonary exercise testing. *Pulm. Circ.* **2016**, *6*, 55–62. [CrossRef] [PubMed]
4. Kell, D.B.; Laubscher, G.J.; Pretorius, E. A central role for amyloid fibrin microclots in long COVID/PASC: Origins and therapeutic implications. *Biochem. J.* **2022**, *479*, 537–559. [CrossRef] [PubMed]

5. Ryu, J.K.; Yan, Z.; Montano, M.; Sozmen, E.G.; Dixit, K.; Suryawanshi, R.K.; Matsui, Y.; Helmy, E.; Kaushal, P.; Makanani, S.K.; et al. Fibrin drives thromboinflammation and neuropathology in COVID-19. *Nature* **2024**, *633*, 905–913. [CrossRef] [PubMed]
6. NIH Opens Long COVID Trials to Evaluate Treatments for Autonomic Nervous System Dysfunction. 12 March 2024. Available online: https://www.nih.gov/news-events/news-releases/nih-opens-long-covid-trials-evaluate-treatments-autonomic-nervous-system-dysfunction (accessed on 9 October 2024).

Journal of *Personalized Medicine*

MDPI

Perspective

Linking the Extended Autonomic System with the Homeostat Theory: New Perspectives about Dysautonomias

David S. Goldstein

Autonomic Medicine Section, Clinical Neurosciences Program, Division of Intramural Research, National Institute of Neurological Disorders and Stroke, National Institutes of Health, Bethesda, MD 20892, USA; goldsteind@ninds.nih.gov; Tel.: +1-301-675-1110; Fax: +1-301-402-0180

Abstract: Dysautonomias are conditions in which altered functions of one or more components of the autonomic nervous system (ANS) adversely affect health. This essay is about how elucidating mechanisms of dysautonomias may rationalize personalized treatments. Emphasized here are two relatively new ideas—the "extended" autonomic system (EAS) and the "homeostat" theory as applied to the pathophysiology and potential treatments of dysautonomias. The recently promulgated concept of the EAS updates Langley's ANS to include neuroendocrine, immune/inflammatory, and central components. The homeostat theory builds on Cannon's theory of homeostasis by proposing the existence of comparators (e.g., a thermostat, glucostat, carbistat, barostat) that receive information about regulated variables (e.g., core temperature, blood glucose, blood gases, delivery of blood to the brain). Homeostats sense discrepancies between the information and response algorithms. The presentation links the EAS with the homeostat theory to understand pathophysiological mechanisms of dysautonomias. Feed-forward anticipatory processes shift input–output curves and maintain plateau levels of regulated variables within different bounds of values—"allostasis". Sustained allostatic processes increase long-term wear-and-tear on effectors and organs—allostatic load. They decreaseing thresholds for destabilizing and potentially fatal positive feedback loops. The homeostat theory enables mathematical models that define stress, allostasis, and allostatic load. The present discussion applies the EAS and homeostat concepts to specific examples of pediatric, adolescent/adult, and geriatric dysautonomias—familial dysautonomia, chronic orthostatic intolerance, and Lewy body diseases. Computer modeling has the potential to take into account the complexity and dynamics of allostatic processes and may yield testable predictions about individualized treatments and outcomes.

Keywords: autonomic; dysautonomia; homeostat; homeostasis; allostasis

1. Introduction

Dysautonomias are conditions in which altered functions of one or more components of the autonomic nervous system (ANS) adversely affect health.

These disorders are frustrating, not only for patients but also for clinicians and researchers. There are several reasons for this. First, dysautonomias come in many forms—there is a whole "universe" of dysautonomias—that can involve essentially all body organs and systems. Because of this multiplicity and the multi-system and therefore multi-disciplinary nature of dysautonomias, they fall through the cracks of the traditional biomedical enterprise. Second, dysautonomias are complex, involving abnormalities in regulation of many effectors and organs by numerous brain anatomic and neurochemical networks. Third, dysautonomias seem often to be mind-body disorders that entail two-way miscommunications between the central autonomic network and body organs; this perspective flies in the face of the traditional Cartesian duality separating the psyche and soma. Fourth, different centers offer diverse autonomic function tests, with the repertoires seeming to heavily depend importantly on cost and throughput, insurance coverage, and regulatory constraints as opposed to tailoring the testing based on relevance to the assessment of individual patients. Fifth, and probably most significant, compared to the large and seemingly increasing patient demand and public health burden, clinical and basic training and scientific knowledge about dysautonomias are disproportionately sparse. This is the "grand challenge" of autonomic disorders [1].

The overall goal of this presentation is to inform the conversation about mechanisms of dysautonomias that may rationalize personalized treatments. Given the above difficulties in the field, one embarks on this sort of essay with some trepidation. Emphasized here are two relatively new ideas—the "extended" autonomic system (EAS) [2] and the "homeostat" theory [3] as applied to the pathophysiology and potential treatments of dysautonomias. I will be considering examples from "galaxies" in the dysautonomias universe, corresponding to pediatric, adult, and geriatric disorders.

2. The "Extended" Autonomic System (EAS)

By mediating automatic, unconscious, involuntary behaviors, the ANS operates at the border of the body and mind. More than a century ago, the English physiologist John Newport Langley defined the ANS as consisting of three parts—the sympathetic nervous system, the parasympathetic nervous system (a phrase he coined), and the enteric nervous system [4]. These were thought to be purely efferent systems for transmitting neuronal signals via ganglia to body organs.

In the intervening century, three types of discoveries have rendered inadequate Langley's theory of the ANS. First, in addition to neurotransmitter systems, a large number of endocrine and neuroendocrine systems mediate automatic, unconscious, involuntary activities within the body's "inner world" [5]. One may reasonably contend that epinephrine was the first hormone and neuroendocrine effector to be identified [6]. Second, ANS components interact complexly and dynamically with immune and inflammatory systems [7]. Third, a brain network that is being described in increasing detail—the central autonomic network [8]—receives and integrates afferent signals from the periphery and modulates autonomic outflows. Based on these considerations, the recently disseminated concept of the "extended" autonomic system (EAS) expands on Langley's ANS to include neuroendocrine, immune/inflammatory, and central facets (Figure 1).

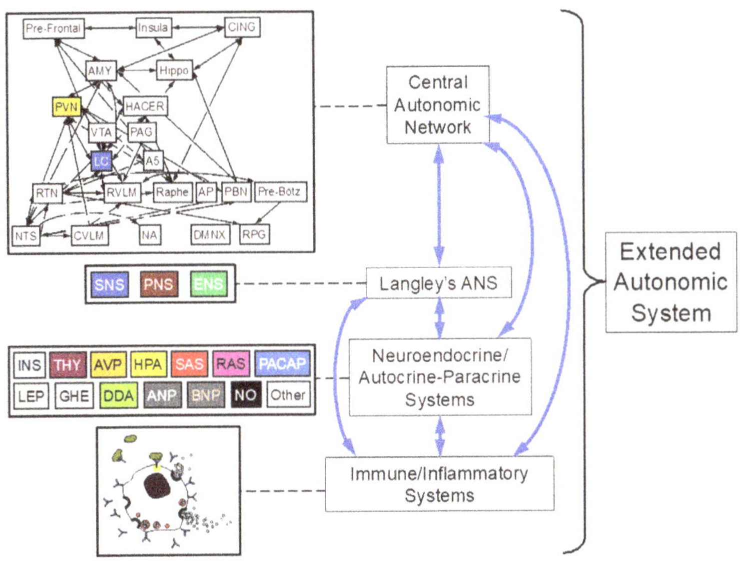

Figure 1. The extended autonomic system. The EAS consists of the central autonomic network (CAN)/stress system, Langley's autonomic nervous system (ANS), neuroendocrine systems, and

inflammatory/immune systems. The four components are bi-directionally inter-related, meaning 6 combinations of relationships. In the central autonomic network, the "stress system" in the Chrousos/Gold schema includes the paraventricular nucleus of the hypothalamus (PVN), which is the source of arginine vasopressin (AVP) and corticotrophin-releasing hormone that drive the hypothalamic-pituitary-adrenocortical axis (HPA), and the pontine locus ceruleus (LC), the main source of norepinephrine in the brain. Abbreviations: Abbreviations: A5 = A5 noradrenergic cell group; AMY = amygdala; ANP = atrial natriuretic peptide; ANS = autonomic nervous system; AP = area posterma; BNP = brain-derived neurotrophic factor; CING = cingulate cortex; CVLM = caudal ventrolateral medulla; DDA = DOPA-dopamine autocrine-paracrine system; DMNX = dorsal motor nucleus of the vagus nerve; ENS = enteric nervous system; GHE = ghrelin; HACER = hypothalamic area controlling emotional responses; Hippo = hippocampus; INS = insulin; LEP = leptin; NA = nucleua ambiguus; NO = nitric oxide; NTS = nuclear of the solitary tract; PACAP = pituitary adenyl cyclase-activating polypeptide; PAG = periaqueductal grey region; PNS = parasympathetic nervous system; Pre-Bötz = pre-Bötzinger complex; RAS = renin-angiotensin-aldosterone system; RPG = respiratory pattern generator; RTN = retrotrapezoid nucleus; RVLM = rostral ventrolateral medulla; SAS = sympathetic adrenergic system; SNS = sympathetic noradrenergic system; THY = thyroid; VTA = ventral tegmental area.

3. The Homeostat Theory

Claude Bernard and Walter B. Cannon (who coined the term "homeostasis") conceptualized that the overall "purpose" of body processes is to maintain the constancy of the internal environment. In contrast, in systems biology, homeostasis is more of an outcome than a goal [9].

The homeostat theory builds on Bernard's and Cannon's notions by proposing the existence of monitored, regulated variables (e.g., core temperature, blood glucose, blood gases, delivery of blood to the brain), which are controlled by comparator "homeostats" (e.g., thermostat, glucostat, carbistat, barostat) that sense discrepancies between afferent information and set points for responding (Figure 2).

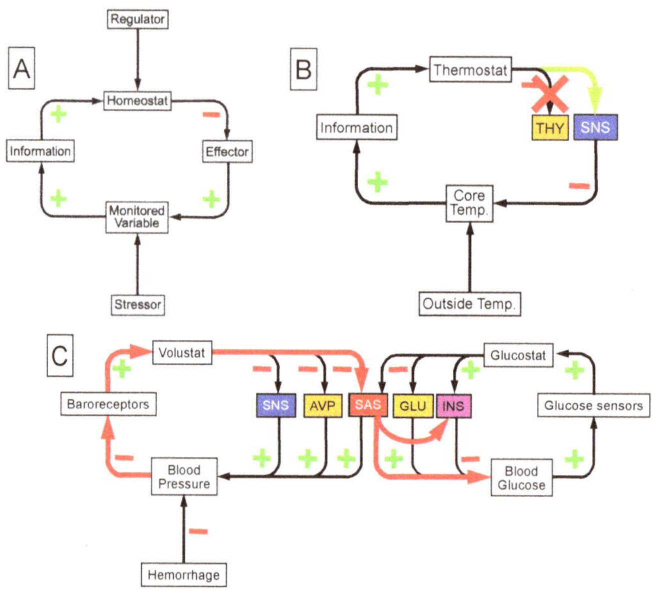

Figure 2. Principles of homeostat operation. Levels of the monitored variable are kept within bounds by negative feedback regulation (**A**). Negative feedback loops are characterized by an odd number of

inhibitory processes (red (−) signs and arrows). Positive relationships are denoted by green + signs. A homeostatic comparator (homeostat) compares afferent information with a set point or other algorithm for responding. The discrepancy drives one or more effectors. (**B**) An example of compensatory activation when there are multiple effectors and one is disabled. Both the hypothalamic-pituitary-thyroid axis and the sympathetic noradrenergic system (SNS) are effectors for regulating core temperature (Core Temp.). Disruption of the hypothalamic-pituitary-thyroid axis compensatorily activates the SNS. (**C**) The sympathetic adrenergic system (SAS) is an effector that is shared by the barostat and glucostat. Effector sharing explains hyperglycemia in hemorrhagic shock. Abbreviations: AVP = arginine vasopressin; GLU = glucagon; INS = insulin.

Homeostats are metaphorical constructs [9]. No one knows what the "purposes" or "goals" of homeostatic systems are. One can postulate the existence of numerous homeostats—an "osmostat" for serum osmolality, an "oxistat" for blood oxygen tension, a "volustat" for effective circulating blood volume, and even a "nocistat" for the experience of pain and a "psychostat" for the sense of equanimity vs. distress. The thought process is that if body variables are kept within bounds, there must be systems at play that are designed to achieve these goals. Systems biologic approaches seem to avoid flirting with this sort of teleological assertion.

4. Allostasis and Allostatic Load

Much of integrative physiological research has focused on negative feedback regulation—reflexes. In humans, however, long-term homeostasis is importantly maintained via anticipatory, feed-forward processes [10] that temporarily shift input–output curves and bring levels of regulated variables to different values—"allostasis" [11].

At first glance, the notion that allostatic processes operate in anticipation of need would seem paradoxical. How can a response occur before the stimulus that would generate that response? This has been a basis for criticizing teleological thinking. Actually, allostatic adjustments can be explained readily by effects of instinct, imprinting, conditioning (both classical (Pavlovian) and operant (instrumental)), and conscious simulations. Examples of instinct in the operations of the EAS would be a person's heart rate increasing as part of "central command" in anticipation of exercise, instinctive avoidance behavior evoked by visual [12] or olfactory [13] predator cues, and innate immune responses to a viral infection. An example of classical conditioning would be augmented tachycardia in anticipation of standing up in patients with postural tachycardia syndrome (POTS) [14] because of learned associations of previously neutral cues with unconditioned aversive stimuli, such as nausea, chest pain, and faintness evoked by orthostasis. An example of operant conditioning would be learning to avoid situations involving prolonged standing, because they are aversive. An example of reacting to conscious simulations would be eating an energy bar before running a mile. Recent animal experiments have begun to identify the specific central pathways and neurochemicals in these responses. In general, they correspond to components of the central autonomic network, although the boundaries of that network seem to require extension to the motor cortex [15] and nigrostriatal dopaminergic system [16].

One of the characteristic features of a viral illness such as COVID-19 is a low-grade fever. According to the allostasis concept, the fever is the result of adjustments in input–output curves for the sympathetic noradrenergic system (SNS), which regulates delivery of blood to the skin surface, and the sympathetic cholinergic system (SCS), which regulates sweating. These adjustments keep core temperature within bounds ("stasis") but at a different level ("allo"). The EAS idea accounts for these allostatic adjustments resulting in fever in COVID-19, in that the EAS incorporates the immune/inflammatory systems and input to the brain from biochemical signals arising from those systems [3].

Allostatic adjustments ordinarily are temporary. For instance, after a viral infection is over, the low-grade fever dissipates. An integrative physiological explanation for dysautonomias is that the allostatic adjustments persist [3]. Levels of regulated variables are kept at new values. This comes at the costs of greater energy utilization, increased variability, and accelerated wear-and-tear on effectors and body organs (allostatic load).

The homeostat theory offers the ability to define difficult entities such as stress and allostatic load in ways that can be modeled mathematically [17,18]. For instance, stress can be defined as the condition in which an error signal drives effectors that decrease the error signal, and allostatic load can be defined as the integrated wear-and-tear on the effectors and consequently on body organs. Among other things, this model predicts that stress can accelerate the accumulation of allostatic load sufficiently to precipitate positive feedback loops and organ failure.

5. Principles of Homeostat Operation

5.1. Multiple Effectors

Having multiple effectors (Figure 2B) offers obvious survival advantages. These include extending the range of control of the monitored variable, compensatory activation of alternative effectors, and stressor-specific patterning. Cannon's view about how blood glucose is maintained included two opposing effectors, the "sympathico-adrenal" system and the "vago-insular" system [19]. If the "common variation" of the level of glycemia fell below a given value (70 mg% was listed), "sympathico-adrenal" activation would raise the glucose level; if the level of glycemia exceeded a given value (130 mg%), "vago-insular" activation would decrease the glucose level. Because of the opposing effectors, the glucose level would be kept within bounds across a range of common variation.

Compensatory activation of alternative effectors enables at least some degree of control of the level of the monitored variable when another effector is disabled. Examples of compensatory activation abound in physiology and pathophysiology, such as recruitment of accessory neck muscles in asthma attacks and augmentation of sympathetic noradrenergic responses to stress in adrenalectomized individuals [20]. Longer-term forms of compensatory activation exemplify plasticity, such as the development of collateral circulation in the setting of coronary artery blockage and adaptive changes in locomotion in movement disorders.

Having multiple effectors probably also permitted the evolution of patterned responses to different stressors. For instance, cold exposure selectively activates the SNS, while glucoprivation selectively activates the sympathetic adrenergic system (SAS) [21].

5.2. Effector Sharing

Effector sharing occurs when two or more homeostatic systems share the same effector. A classic example is the arginine vasopressin (AVP) system. AVP not only is a vasoconstrictor but also, acting as the anti-diuretic hormone, is the body's main effector in regulation of water balance and hence of serum osmolality. Sharing of the AVP effector by the "barostat" and "osmostat" explains why patients in shock can be hyponatremic. Similarly, sharing of the SAS effector by the barostat and "glucostat" explains why the patients are hyperglycemic (Figure 2C).

6. Homeostats at Work

The key elements of the homeostat theory—monitored variables, regulators, and homeostats—are in essence metaphors. Experimental observations over the last two decades, however, have increasingly elucidated how homeostatic systems operate and have generally supported the concepts of multiple effectors, effector sharing, negative feedback regulation, and allostasis. The following discussion focuses on regulation of core temperature, glucose, blood gases, and delivery of blood to the brain during orthostasis.

6.1. Thermoregulation

Humans have two primary sources of afferent information about temperature, the skin and the arterial blood (Figure 3). A neuronal pathway relays cutaneous sensory information via the dorsal horn, spinothalamic tract, and lateral brachial nucleus to the pre-optic area

(POA), which also possesses neurons responsive to the temperature of the arterial blood. Subjective thermal comfort plays a critical role in body temperature regulation, since this represents the primary stimulus for behavioral thermoregulation. Core and skin temperature contribute about equally to thermal comfort, whereas metabolic heat production and plasma catecholamine responses are more responsive to core temperature [22].

In fruit flies, peripheral thermosensory information to higher brain centers converges onto three target regions: the mushroom body, the lateral horn, and the posterior lateral protocerebrum. Hot and cold antennal receptors project onto distinct but adjacent glomeruli in the proximal antennal protocerebrum, forming a thermotopic map in the brain. It has been proposed that "... dedicated populations of cells orchestrate behavioral responses to different temperature stimuli, and reveal a labeled-line logic for the coding of temperature information in the brain" [23].

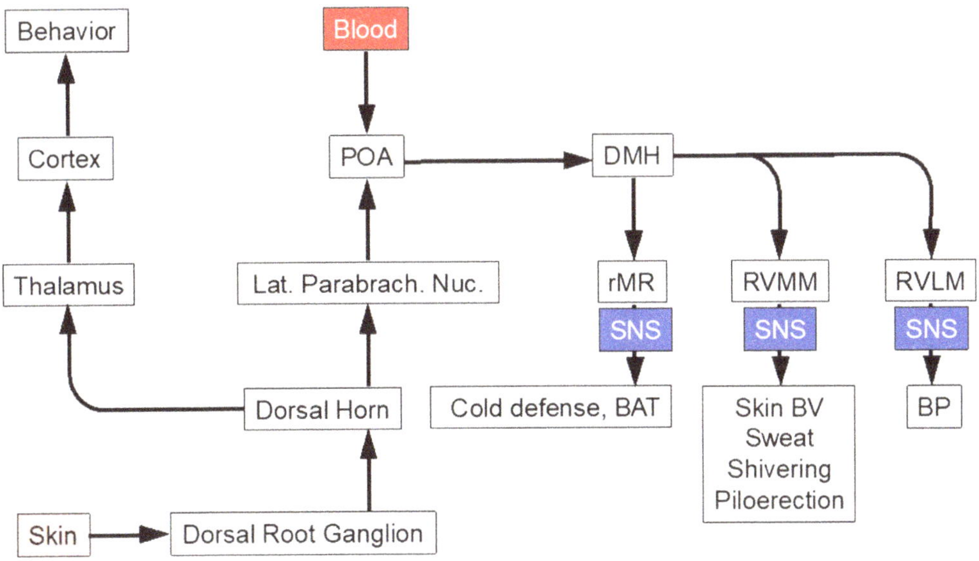

Figure 3. Central network controlling core temperature via the sympathetic noradrenergic system (SNS). Abbreviations: BAT = brown adipose tissue; BP = blood pressure; BV = blood vessels; DMH = dorsomedial hypothalamus; POA = pre-optic area; rMR = rostral medullary raphe; RVLM = rostral ventrolateral medulla; RVMM = rostral ventromedial medulla.

6.2. Glucose

The hypothalamic arcuate nucleus (ARC) is a key brain center for sensing adiposity signals (e.g., insulin, leptin, ghrelin, glucagon-related peptide 1) (Figure 4). ARC neurons not only regulate feeding but also contribute to glucose homeostasis and innate immune responses.

Blood levels of glucose are regulated mainly by hormones, such as insulin from pancreatic islet β-cells, glucagon from pancreatic islet α-cells, epinephrine from adrenomedullary chromaffin cells, and, to a lesser extent, cortisol from adrenocortical zona fasciculata cells. These hormonal effects interact complexly. Glucagon may increase circulating glucose levels both directly via hepatic glucose release and indirectly via adrenomedullary epinephrine secretion [24]. Meanwhile, epinephrine stimulates pancreatic glucagon secretion [25], suggesting the potential for a self-reinforcing positive feedback loop. Epinephrine stimulates pancreatic insulin secretion via β-adrenoceptors but mainly inhibits insulin secretion via agonism at α-adrenoceptors. Epinephrine infusion blunts insulin responses to both hyperglycemia and glucagon [26].

Virtually every serious illness or cause of emotional distress is associated with hyperglycemia, even in individuals without a history of diabetes, and is associated with worse outcome [27–31]. One may reasonably propose that the adverse prognoses associated with hyperglycemia are not the result of hyperglycemia itself so much as of disease severity-related neuroendocrine changes producing hyperglycemia.

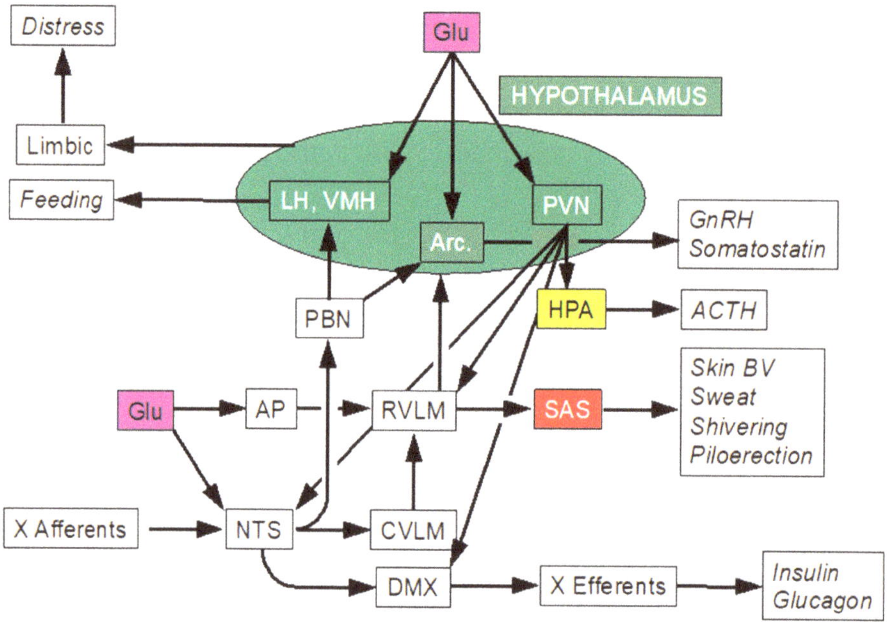

Figure 4. Central network controlling blood glucose. Abbrevations: AP = area postrema; Arc. = arcuate nucleus; BV = blood vessels; CVLM = caudal ventrolateral medulla; DMX = dorsal motor nucleus of the vagus nerve; GnRH = growth hormone-releasing hormone; HPA = hypothalamic-pituitary-adrenocortical axis; LH = lateral hypothalamus; NTS = nucleus of the solitary tract; PBN = parabrachial nucleus; PVN = paraventricular nucleus of the hypothalamus; RVLM = rostral ventrolateral medulla; SAS = sympathetic adrenergic system; VMH = ventromedial hypothalamic nucleus; X = vagus nerve.

6.3. Blood Gases

In mammals, appropriate delivery of oxygen to and removal of carbon dioxide are crucial for survival. Multiple effectors for this regulation exist, and blocking one compensatorily activates others. The retrotrapezoid nucleus (RTN) neurons in the rostral ventrolateral medulla (RVLM) is part of a column of respiration-related neuronal clusters. The RTN is thought to regulate breathing automaticity and arterial pCO_2 homeostasis (Figure 5). The carotid bodies stimulate the respiratory pattern generator both directly and indirectly by activating the RTN via a neuronal projection originating within the nucleus of the solitary tract (NTS). Consistent with the principle of multiple effects and compensatory activation, silencing RTN neurons increases carotid body activity [32].

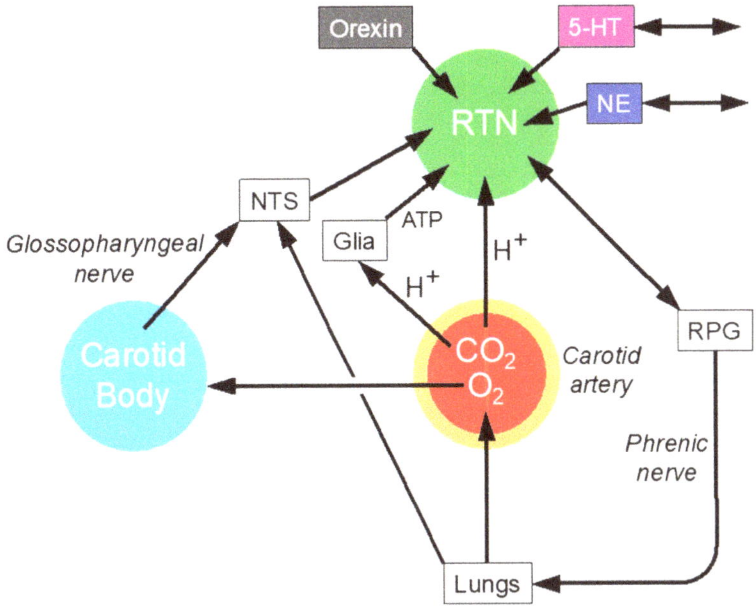

Figure 5. Central network controlling arterial blood gases.

6.4. Blood Flow to the Brain during Orthostasis

The requirements of correct core temperature and continuous availability of metabolic fuels have challenged organismic integrity throughout mammalian evolution. Multiple effectors have evolved to meet these challenges. In contrast, humans have been standing up since only relatively recently in evolutionary time. It is thought that in Africa, about 5–6 million years ago, there was a shift from jungle to savannah life. Bipedalism afforded obvious selective advantages in this new ecological niche, such as seeing further distances during migrating, carrying objects and infants, communication via hand gestures or arm waving, and more powerful striking and manipulating. According to cladographic data, our ancestor *Homo erectus* came on the scene only about 2–3 million years ago.

In order to tolerate standing, an individual must be able to tighten blood vessels below the level of the heart and increase the force and rate of cardiac contraction to maintain blood flow to the brain. One may speculate that because orthostasis is relatively new in evolutionary terms, only one system, the SNS, is available to maintain blood flow to the brain during orthostasis. Predictably, orthostatic intolerance and hypotension are cardinal manifestations of SNS failure.

There are two general types of afferent information to the brain during orthostasis (Figure 6). The first is high-pressure mechanoreceptors in the walls of arteries—especially in the carotid sinus, at the vascular gateway to the brain. The second is low-pressure mechanoreceptors in atria and pulmonary veins. Both types of mechanoreceptors are unloaded by the orthostatic decrease in venous return to the heart.

The effectors mediating the homeostatic responses are similar, but there are some differences. Activation of the renin-angiotensin-aldosterone system seems to be more prominent with unloading of low- than of high-pressure mechanoreceptors [33]. Low-pressure mechanoreceptors also appear to play a prominent role in reflexive forearm vasoconstriction [34] and SAS activation [35].

Lower body negative pressure (LBNP) decreases venous return to the heart and simulates gravitational stress. Reflexive sympathetically-mediated vasoconstriction can explain maintenance of arterial blood pressure in this setting. Non-hypotensive LBNP decreases middle cerebral artery blood flow velocity without a change in arterial diameter [36]. These findings indicate that during orthostasis, brain blood flow decreases for the same level of blood pressure—operationally, an allostatic shift in the chair-shaped curve relating cerebral blood flow to blood pressure (autoregulation). One may hypothesize that individuals with relatively large orthostatic decreases in venous return to the heart would be more likely to have symptoms of orthostatic intolerance, such as lightheadedness or "brain fog". Testing this hypothesis would require controlling for hyperventilation, which independently decreases middle cerebral artery blood flow velocity by decreasing arterial pCO_2.

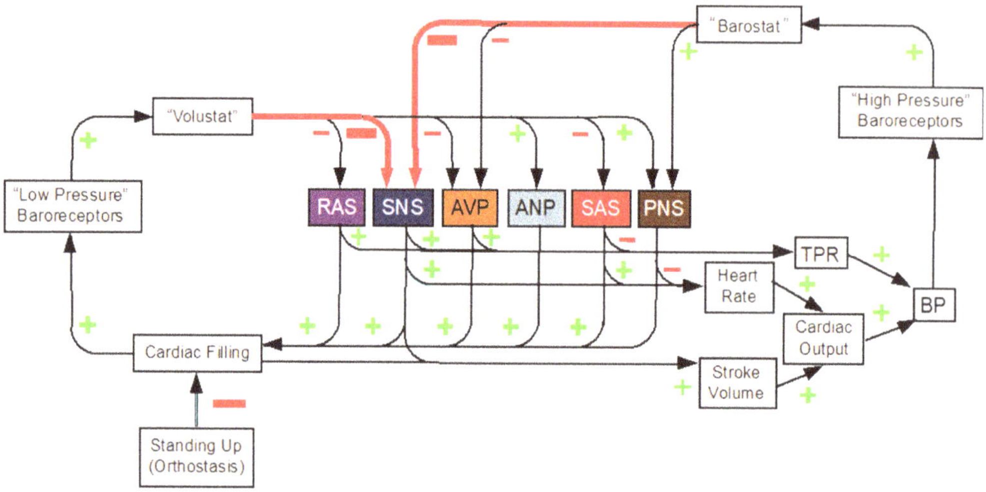

Figure 6. Low- and high-pressure baroreflexes. The sympathetic noradrenergic system (SNS) is the main effector for maintaining blood pressure (BP) during orthostasis. Diffuse SNS failure always manifests with orthostatic hypotension. Other effectors are the renin-angiotensin-aldosterone system (RAS), arginine vasopressin system (AVP), atrial natriuretic peptide (ANP), the sympathetic adrenergic system (SAS), and the parasympathetic nervous system (PNS).

7. Application to Pediatric Dysautonomias: Familial Dysautonomia (FD)

Within the dysautonomias universe, the pediatric "galaxy" often entails substantial genetic load or embryological abnormalities in development of components of the ANS.

A classic example is familial dysautonomia (FD), also referred to as Riley-Day syndrome and Type III hereditary sensory and autonomic neuropathy (HSAN III). FD is mainly a disease of people of Ashkenazic Jewish extraction, due to a founder effect; almost all the disease alleles share a common ancestral haplotype. The disease results from a splicing error in the Elongator acetyltransferase complex subunit 1 (*ELP1*) gene (also known as *IKBKAP*). The splicing error results in exon 20 being skipped in different tissues.

The pattern of plasma levels of catechols in FD points to arrested development of sympathetic noradrenergic nerves, coupled with compensatorily increased activity of tyrosine hydroxylase and normal activity of the SAS [37]. FD patients have attenuated orthostatic increments in plasma norepinephrine levels [38], possibly reflecting a generalized abnormality of sensory afferents, including from mechanoreceptors [39].

FD patients are susceptible to crises of nausea and vomiting associated with tachycardia, sweating, hypertension, and behavioral changes. Cyclic vomiting in FD is associated with high circulating dopamine levels [40]. This hyperdopaminergic state seems to be pathophysiologically significant, because treatment with carbidopa, which inhibits catecholamine biosynthesis, is effective in mitigating the vomiting [41]. Vesicles containing newly synthesized norepinephrine are released preferentially during sympathetic stimulation [42], and acute increases in plasma dopamine are likely to reflect increased exocytotic release from sympathetic noradrenergic nerves. It is therefore reasonable to speculate that arrested development of sympathetic noradrenergic nerves in FD results in a form of functional dopamine-beta-hydroxylase deficiency and compensatorily increasing sympathetic traffic to extant terminals, so that during crises there is excessive dopamine release compared to the increases in plasma levels of norepinephrine and epinephrine.

Multi-disciplinary management strategies have improved survival in FD. Experimental therapeutic efforts to treat the disease process itself have so far been unsuccessful. After development of an animal model of FD and high-throughput drug screening, the small molecule kinetin (6-furfurylaminopurine) seemed promising. The pharmaceutical development program ended in 2019 due to budgetary constraints and the rarity of the patient population. Other feasible therapeutic approaches are small nuclear RNA components [43] or antisense oligonucleotides [44] to treat the splicing defect. Also, gene replacement therapy has been proposed that would entail delivering Type 2 adeno-associated virus (AAV) to express a wild type copy of the *ELP1* gene [45] or Type 9 AAV for exon-specific inclusion of *ELP1* exon 20 in cells expressing the target pre-mRNA [46].

The most effective treatment for FD would be prevention of the disease. An effort is under way to avoid reproduction by heterozygous carriers [47]; theoretically, this might eventually eliminate the disease.

8. Application to Adult Dysautonomias: Postural Tachycardia Syndrome (POTS)

Dysautonomias in adolescents or adults often involve complex, multi-system disorders of regulation of components of the ANS, where the effectors have developed normally. Chronic orthostatic intolerance in POTS and repeated episodes of neurocardiogenic syncope (NCS) involve many symptoms, such as fatigue, exercise and heat intolerance, presyncope, impaired concentration and memory, headache, coat hanger pain, early satiety, bloating or vomiting, tremulousness, and pallor.

Both POTS and NCS are far more common in women than men, for reasons that remain poorly understood. Among vigorously healthy astronauts re-exposed to the earth's gravity after prolonged space flight, orthostasis intolerance is far more prevalent in females. Application of a computer model of cardiovascular function has indicated that simple differences in physiognomy such as the longitudinal center of gravity can explain the greater prevalence of post-reentry orthostatic intolerance in women than men [48]. For the same orthostatic gravitational stress, women might have a greater shift in blood volume to pelvic veins and therefore a larger fall in venous return to the heart and cardiac stroke volume [49].

The schema in Figure 7 offers a concept for how neurocirculatory dyshomeostasis might result in persistent fatigue, a tendency to faint, excessive orthostatic tachycardia, and brain fog in POTS. The red arrows indicate afferent input to the central autonomic network from "high pressure" arterial baroreceptors that respond to alterations in systemic blood pressure, "low pressure" baroreceptors that respond to alterations in pulmonary venous pressure, and signals from the immune/inflammatory system. The numerous inter-relationships, most of which are bi-directional, seem dauntingly complex, yet they are derived from two relatively simple ideas, the EAS and the homeostat theory.

In general, chronic orthostatic intolerance syndromes do not evolve to lethal neurodegenerative diseases, and in a substantial proportion of cases, the overall clinical status improves over time. Therapeutic interventions in which patients actively participate, such as graded exercise or counter-maneuvers [50], meditation, or yoga [51], might improve symptoms because of SNS activation in the setting of active coping [52].

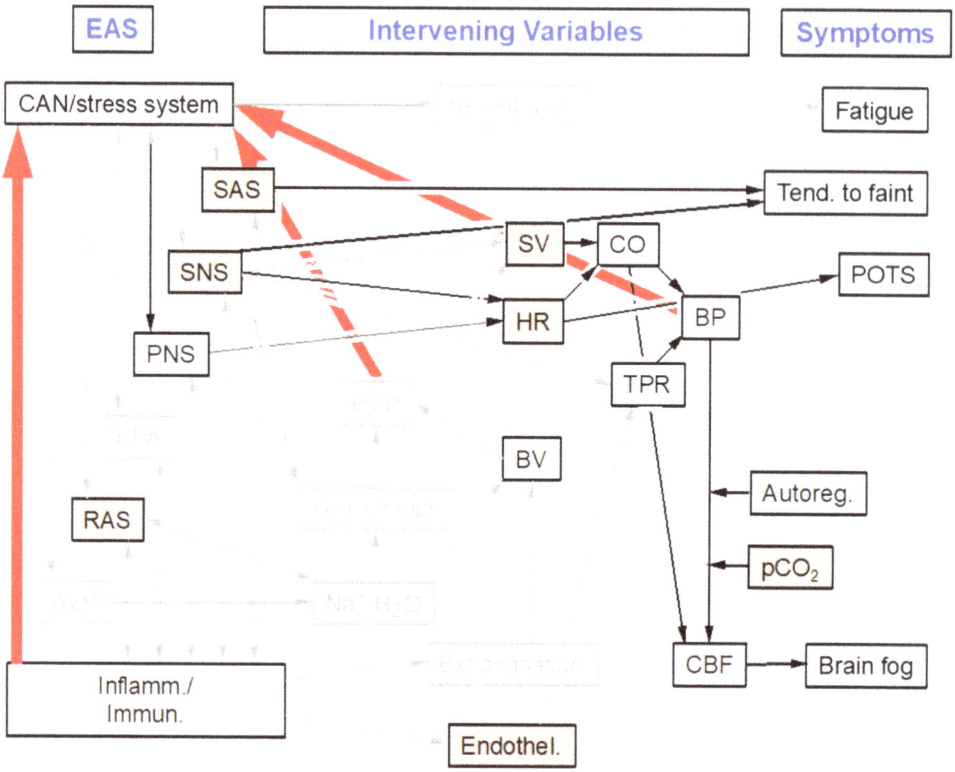

Figure 7. Concept diagram relating the EAS to intervening variables to symptoms of brain fog, a tendency to faint, and orthostatic intolerance in post-infectious POTS. Red arrows indicate afferent input to the brain from high-pressure and low-pressure mechanoreceptors. Grayed out boxes indicate variables for which objective data in POTS are incomplete or inconsistent. Pink filling indicates variables with abnormal values in POTS. Imbalance between sympathetic noradrenergic system (SNS) and sympathetic adrenergic system (SAS) outflows produces a tendency to faint. Other abbreviations: AVP = arginine vasopressin; Autoreg. = cerebrovascular autoregulation; BV = blood volume; BP = arterial blood pressure; CBF = cerebral blood flow; Endothel. = endothelial dysfunction; Neurotransm. = central neurotransmitters; POTS = postural tachycardia syndrome; PVP = pulmonary venous pressure; RAS = renin-angiotensin-aldosterone; SV = cardiac stroke volume; Tend. to faint = tendency to faint; TPR = total peripheral vascular resistance; Ven. Compl. = splanchnic venous compliance.

9. Application to Geriatric Dysautonomias: Central Lewy Body Diseases (LBDs)

A major form of geriatric dysautonomias is a family of diseases involving Lewy bodies, intra-neuronal inclusion bodies having characteristic histopathological features. In Lewy body diseases (LBDs), Lewy bodies are found in brainstem dopaminergic and noradrenergic neurons or in sympathetic ganglia. Lewy bodies contain an abundance of the protein alpha-synuclein (αS). Conditions previously classified as forms of primary chronic autonomic failure—pure autonomic failure (PAF), multiple system atrophy (MSA), and Parkinson's disease with orthostatic hypotension (PD + OH)—are referred to as autonomic synucleinopathies [53]. Dementia with Lewy bodies (DLB) involves a relatively high frequency of orthostatic hypotension and neuroimaging evidence of cardiac noradrenergic deficiency [54] and is now included in the family of autonomic synucleinopathies. All these disorders involve catecholamine deficiencies in the brain, the periphery, or both.

In the central LBDs PD and DLB, by the time parkinsonism or cognitive dysfunction manifests, clinically substantial catecholaminergic neurodegeneration has already occurred. Neurorescue strategies might forestall symptomatic disease if central LBDs could be identified in a preclinical phase. The prospective, observational, long-term PDRisk study assessed the predictive value of low vs. normal cardiac ^{18}F-dopamine positron emission tomography (PET), an index of myocardial content of the sympathetic neurotransmitter norepinephrine [55] in at-risk individuals. At 7 years of follow-up, eight of nine participants with low initial ^{18}F-dopamine-derived radioactivity and one of eleven with normal radioactivity were subsequently diagnosed with a central LBD (LBD+). Conversely, all of nine LBD+ participants had low radioactivity before or at the time of diagnosis of a central LBD, whereas among twenty-five participants without a central LBD, only one (4%) had persistently low radioactivity. Cardiac ^{18}F-dopamine PET therefore highly efficiently distinguishes at-risk individuals who are subsequently diagnosed with a central LBD from those who are not [56]. These results have supported the view that the pathophysiological process leading to central LBDs can begin outside the brain, with early involvement of the autonomic nervous system—especially sympathetic noradrenergic innervation of the heart [55].

Computational modeling has revealed multiple functional abnormalities in catecholaminergic neurons in LBDs [57]. These abnormalities can be explained by autotoxic interactions between oxidized metabolites of catecholamines and αS [58]. Extension of the modeling to address the trajectory of loss of catecholamine stores in LBDs over time has indicated a tri-phasic pattern [59] (Figure 8). For years, compensatory activation maintains homeostasis of striatal dopamine [60]. Once the compensatory processes are overwhelmed because of autotoxicity and allostatic load producing aging-related declines in efficiency, a second phase ensues in which there is a rapid decline in neurotransmitter stores (dyshomeostasis). When the complement of releasable catecholamine falls below a threshold level, the patient notes symptoms of the deficiency. In the symptomatic third phase, there is slow further loss.

The key to delaying the onset of symptomatic catecholaminergic neurodegeneration would be to begin treatment soon after the transition from homeostasis to dyshomeostasis. Mathematical modeling predicts that the same treatment that would exert only a small, transient benefit in symptomatic disease, but begun at the transition from homeostasis to dyshomeostasis, would substantially delay the onset of symptomatic disease [59].

It seems reasonable to propose that computational modeling, coupled with empirical data about EAS effectors and intervening variables, might yield testable hypotheses about exacerbating/ameliorating factors, responses to treatments, and outcomes in individual patients. Such a project, however, would require coordinating the efforts of integrative physiologists, systems biologists, and autonomic neuroscientists.

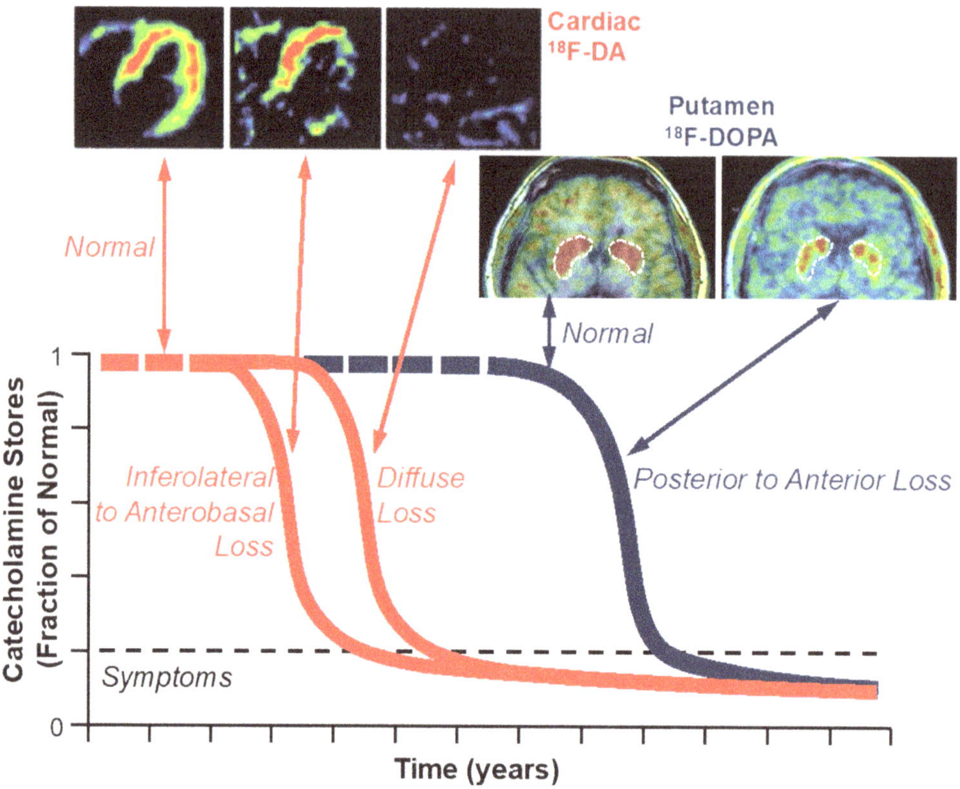

Figure 8. Tri-phasic loss of catecholamine stores in Lewy body diseases. Cardiac noradrenergic stores assessed by ^{18}F-dopamine (^{18}F-DA) positron emission tomography decline in a tri-phasic manner before tri-phasic decline in putamen ^{18}F-DOPA-derived radioactivity. The loss of left ventricular myocardial ^{18}F-DA-derived radioactivity proceeds from the inferolateral to the anterobasal wall, and the loss of putamen ^{18}F-DOPA-derived radioactivity proceeds from the posterior to the anterior putamen.

10. Conclusions

The EAS expands on the ANS by including neuroendocrine systems, immune/inflammatory systems, and the central autonomic network. The four components interact complexly and bi-directionally and determine clinical manifestations of dysautonomias. The homeostat theory enables objective, non-circular definitions of stress, allostasis, and allostatic load. Computer modeling has the potential to take into account the complexity and dynamics of allostatic processes [18,61] and may yield testable predictions about individualized treatments and outcomes.

Funding: Goldstein was supported (in part) by the Division of Intramural Research of the NIH, NINDS.

Institutional Review Board Statement: Not applicable.

Informed Consent Statement: Not applicable.

Data Availability Statement: No new data were created or analyzed in this study. Data sharing is not applicable to this article.

Acknowledgments: The content is solely the responsibility of the author and does not necessarily represent the official views of the National Institutes of Health.

Conflicts of Interest: The author declares no conflict of interest

References

1. Cheshire, W.P. The grand challenge of autonomic disorders. *Front. Neurol.* **2022**, *13*, 1052137. [CrossRef] [PubMed]
2. Goldstein, D.S. The extended autonomic system, dyshomeostasis, and COVID-19. *Clin. Auton. Res.* **2020**, *30*, 299–315. [CrossRef] [PubMed]
3. Goldstein, D.S. Stress and the "extended" autonomic system. *Auton. Neurosci.* **2021**, *236*, 102889. [CrossRef]
4. Langley, J.N. The autonomic nervous system. *Brain* **1903**, *26*, 1–26. [CrossRef]
5. Goldstein, D.S. *Adrenaline and the Inner World: An Introduction to Scientific Integrative Medicine*; The Johns Hopkins University Press: Baltimore, MD, USA, 2006.
6. Rao, Y. The First Hormone: Adrenaline. *Trends Endocrinol. Metab.* **2019**, *30*, 331–334. [CrossRef] [PubMed]
7. Udit, S.; Blake, K.; Chiu, I.M. Somatosensory and autonomic neuronal regulation of the immune response. *Nat. Rev. Neurosci.* **2022**, *23*, 157–171. [CrossRef]
8. Benarroch, E.E. Physiology and Pathophysiology of the Autonomic Nervous System. *Continuum* **2020**, *26*, 12–24. [CrossRef]
9. Goldstein, D.S. How does homeostasis happen? Integrative physiological, systems biological, and evolutionary perspectives. *Am. J. Physiol. Regul. Integr. Comp. Physiol.* **2019**, *316*, R301–R317. [CrossRef]
10. Schulkin, J.; Sterling, P. Allostasis: A Brain-Centered, Predictive Mode of Physiological Regulation. *Trends Neurosci.* **2019**, *42*, 740–752. [CrossRef]
11. Sterling, P. *What Is Health?: Allostasis and the Evolution of Human Design*; MIT Press: Cambridge, MA, USA, 2020.
12. Li, A.J.; Wang, Q.; Ritter, S. Selective Pharmacogenetic Activation of Catecholamine Subgroups in the Ventrolateral Medulla Elicits Key Glucoregulatory Responses. *Endocrinology* **2018**, *159*, 341–355. [CrossRef]
13. Kondoh, K.; Lu, Z.; Ye, X.; Olson, D.P.; Lowell, B.B.; Buck, L.B. A specific area of olfactory cortex involved in stress hormone responses to predator odours. *Nature* **2016**, *532*, 103–106. [CrossRef] [PubMed]
14. Norcliffe-Kaufmann, L.; Palma, J.A.; Martinez, J.; Camargo, C.; Kaufmann, H. Fear conditioning as a pathogenic mechanism in the postural tachycardia syndrome. *Brain* **2022**, *145*, 3763–3769. [CrossRef] [PubMed]
15. Strack, A.M.; Sawyer, W.B.; Platt, K.B.; Loewy, A.D. CNS cell groups regulating the sympathetic outflow to adrenal gland as revealed by transneuronal cell body labeling with pseudorabies virus. *Brain Res.* **1989**, *491*, 274–296. [CrossRef] [PubMed]
16. Chen, A.P.F.; Chen, L.; Shi, K.W.; Cheng, E.; Ge, S.; Xiong, Q. Nigrostriatal dopamine modulates the striatal-amygdala pathway in auditory fear conditioning. *Nat. Commun.* **2023**, *14*, 7231. [CrossRef] [PubMed]
17. Goldstein, D.S. Computer models of stress, allostasis, and acute and chronic disease. *Ann. N. Y. Acad. Sci.* **2008**, *1148*, 223–231. [CrossRef] [PubMed]
18. Acevedo, A.; Androulakis, I.P. Allostatic breakdown of cascading homeostat systems: A computational approach. *Heliyon* **2017**, *3*, e00355. [CrossRef] [PubMed]
19. Cannon, W.B. Organization for physiological homeostasis. *Physiol. Rev.* **1929**, *9*, 399–431. [CrossRef]
20. Udelsman, R.; Goldstein, D.S.; Loriaux, D.L.; Chrousos, G.P. Catecholamine-glucocorticoid interactions during surgical stress. *J. Surg. Res.* **1987**, *43*, 539–545. [CrossRef]
21. Goldstein, D.S.; Kopin, I.J. Adrenomedullary, adrenocortical, and sympathoneural responses to stressors: A meta-analysis. *Endo. Regul.* **2008**, *42*, 111–119.
22. Frank, S.M.; Raja, S.N.; Bulcao, C.F.; Goldstein, D.S. Relative contribution of core and cutaneous temperatures to thermal comfort and autonomic responses in humans. *J. Appl. Physiol.* **1999**, *86*, 1588–1593. [CrossRef]
23. Gallio, M.; Ofstad, T.A.; Macpherson, L.J.; Wang, J.W.; Zuker, C.S. The coding of temperature in the Drosophila brain. *Cell* **2011**, *144*, 614–624. [CrossRef] [PubMed]
24. Sharabi, Y.; Goldstein, D.S.; Bentho, O.; Saleem, A.; Pechnik, S.; Geraci, M.F.; Holmes, C.; Pacak, K.; Eisenhofer, G. Sympathoadrenal function in patients with paroxysmal hypertension: Pseudopheochromocytoma. *J. Hypertens.* **2007**, *25*, 2286–2295. [CrossRef] [PubMed]
25. Hamilton, A.; Zhang, Q.; Salehi, A.; Willems, M.; Knudsen, J.G.; Ringgaard, A.K.; Chapman, C.E.; Gonzalez-Alvarez, A.; Surdo, N.C.; Zaccolo, M.; et al. Adrenaline Stimulates Glucagon Secretion by Tpc2-Dependent Ca(2+) Mobilization From Acidic Stores in Pancreatic alpha-Cells. *Diabetes* **2018**, *67*, 1128–1139. [CrossRef] [PubMed]
26. Porte, D., Jr.; Graber, A.L.; Kuzuya, T.; Williams, R.H. The effect of epinephrine on immunoreactive insulin levels in man. *J. Clin. Investig.* **1966**, *45*, 228–236. [CrossRef] [PubMed]
27. Wu, J.; Huang, J.; Zhu, G.; Wang, Q.; Lv, Q.; Huang, Y.; Yu, Y.; Si, X.; Yi, H.; Wang, C.; et al. Elevation of blood glucose level predicts worse outcomes in hospitalized patients with COVID-19: A retrospective cohort study. *BMJ Open Diabetes Res. Care* **2020**, *8*, e001476. [CrossRef] [PubMed]
28. Singh, A.K.; Singh, R. Does poor glucose control increase the severity and mortality in patients with diabetes and COVID-19? *Diabetes Metab. Syndr.* **2020**, *14*, 725–727. [CrossRef]

29. Scheen, A.J.; Marre, M.; Thivolet, C. Prognostic factors in patients with diabetes hospitalized for COVID-19: Findings from the CORONADO study and other recent reports. *Diabetes Metab.* **2020**, *46*, 265–271. [CrossRef] [PubMed]
30. Bode, B.; Garrett, V.; Messler, J.; McFarland, R.; Crowe, J.; Booth, R.; Klonoff, D.C. Glycemic Characteristics and Clinical Outcomes of COVID-19 Patients Hospitalized in the United States. *J. Diabetes Sci. Technol.* **2020**, *14*, 813–821. [CrossRef]
31. Dziewierz, A.; Giszterowicz, D.; Siudak, Z.; Rakowski, T.; Dubiel, J.S.; Dudek, D. Admission glucose level and in-hospital outcomes in diabetic and non-diabetic patients with acute myocardial infarction. *Clin. Res. Cardiol.* **2010**, *99*, 715–721. [CrossRef]
32. Guyenet, P.G.; Bayliss, D.A.; Stornetta, R.L.; Kanbar, R.; Shi, Y.; Holloway, B.B.; Souza, G.; Basting, T.M.; Abbott, S.B.G.; Wenker, I.C. Interdependent feedback regulation of breathing by the carotid bodies and the retrotrapezoid nucleus. *J. Physiol.* **2018**, *596*, 3029–3042. [CrossRef]
33. Egan, B.M.; Julius, S.; Cottier, C.; Osterziel, K.J.; Ibsen, H. Role of cardiovascular receptors on the neural regulation of renin release in normal men. *Hypertension* **1983**, *5*, 779–786. [CrossRef] [PubMed]
34. Zoller, R.P.; Mark, A.L.; Abboud, F.M.; Schmid, P.G.; Heistad, D.D. The role of low pressure baroreceptors in reflex vasoconstrictor responses in man. *J. Clin. Investig.* **1972**, *51*, 2967–2972. [CrossRef] [PubMed]
35. Jacobs, M.C.; Goldstein, D.S.; Willemsen, J.J.; Smits, P.; Thien, T.; Lenders, J.W. Differential effects of low- and high-intensity lower body negative pressure on noradrenaline and adrenaline kinetics in humans. *Clin. Sci.* **1996**, *90*, 337–343. [CrossRef] [PubMed]
36. Serrador, J.M.; Picot, P.A.; Rutt, B.K.; Shoemaker, J.K.; Bondar, R.L. MRI measures of middle cerebral artery diameter in conscious humans during simulated orthostasis. *Stroke* **2000**, *31*, 1672–1678. [CrossRef] [PubMed]
37. Goldstein, D.S.; Holmes, C.; Axelrod, F.B. Plasma catechols in familial dysautonomia: A long-term follow-up study. *Neurochem. Res.* **2008**, *33*, 1889–1893. [CrossRef] [PubMed]
38. Axelrod, F.B.; Goldstein, D.S.; Holmes, C.; Berlin, D.; Kopin, I.J. Pattern of plasma levels of catecholamines in familial dysautonomia. *Clin. Auton. Res.* **1996**, *6*, 205–209. [CrossRef] [PubMed]
39. Norcliffe-Kaufmann, L.; Axelrod, F.; Kaufmann, H. Afferent baroreflex failure in familial dysautonomia. *Neurology* **2010**, *75*, 1904–1911. [CrossRef] [PubMed]
40. Norcliffe-Kaufmann, L.J.; Axelrod, F.B.; Kaufmann, H. Cyclic vomiting associated with excessive dopamine in Riley-day syndrome. *J. Clin. Gastroenterol.* **2013**, *47*, 136–138. [CrossRef]
41. Norcliffe-Kaufmann, L.; Martinez, J.; Axelrod, F.; Kaufmann, H. Hyperdopaminergic crises in familial dysautonomia: A randomized trial of carbidopa. *Neurology* **2013**, *80*, 1611–1617. [CrossRef]
42. Kopin, I.J.; Breese, G.R.; Krauss, K.R.; Weise, V.K. Selective release of newly synthesized norepinephrine from the cat spleen during sympathetic nerve stimulation. *J. Pharmacol. Exp. Ther.* **1968**, *161*, 271–278.
43. Hatch, S.T.; Smargon, A.A.; Yeo, G.W. Engineered U1 snRNAs to modulate alternatively spliced exons. *Methods* **2022**, *205*, 140–148. [CrossRef] [PubMed]
44. Sinha, R.; Kim, Y.J.; Nomakuchi, T.; Sahashi, K.; Hua, Y.; Rigo, F.; Bennett, C.F.; Krainer, A.R. Antisense oligonucleotides correct the familial dysautonomia splicing defect in IKBKAP transgenic mice. *Nucleic Acids Res.* **2018**, *46*, 4833–4844. [CrossRef] [PubMed]
45. Schultz, A.; Cheng, S.Y.; Kirchner, E.; Costello, S.; Miettinen, H.; Chaverra, M.; King, C.; George, L.; Zhao, X.; Narasimhan, J.; et al. Reduction of retinal ganglion cell death in mouse models of familial dysautonomia using AAV-mediated gene therapy and splicing modulators. *Sci. Rep.* **2023**, *13*, 18600. [CrossRef] [PubMed]
46. Romano, G.; Riccardi, F.; Bussani, E.; Vodret, S.; Licastro, D.; Ragone, I.; Ronzitti, G.; Morini, E.; Slaugenhaupt, S.A.; Pagani, F. Rescue of a familial dysautonomia mouse model by AAV9-Exon-specific U1 snRNA. *Am. J. Hum. Genet.* **2022**, *109*, 1534–1548. [CrossRef]
47. Singer, A.; Sagi-Dain, L. Impact of a national genetic carrier-screening program for reproductive purposes. *Acta Obstet. Gynecol. Scand.* **2020**, *99*, 802–808. [CrossRef]
48. Summers, R.L.; Platts, S.; Myers, J.G.; Coleman, T.G. Theoretical analysis of the mechanisms of a gender differentiation in the propensity for orthostatic intolerance after spaceflight. *Theor. Biol. Med. Model.* **2010**, *7*, 8. [CrossRef]
49. Diaz-Canestro, C.; Sehgal, A.; Pentz, B.; Montero, D. Sex specificity in orthostatic tolerance: The integration of haematological, cardiac, and endocrine factors. *Eur. J. Prev. Cardiol.* **2022**, *29*, e246–e248. [CrossRef]
50. Fu, Q.; Levine, B.D. Exercise and non-pharmacological treatment of POTS. *Auton. Neurosci.* **2018**, *215*, 20–27. [CrossRef]
51. Abdelazeem, B.; Abbas, K.S.; Manasrah, N.; Amin, M.A.; Mohammed, S.M.; Mostafa, M.R. Yoga as a treatment for vasovagal syncope: A systematic review and meta-analysis. *Complement. Ther. Clin. Pract.* **2022**, *48*, 101579. [CrossRef]
52. De Boer, S.F.; Slangen, J.L.; Van der Gugten, J. Plasma catecholamine and corticosterone levels during active and passive shock-prod avoidance behavior in rats: Effects of chlordiazepoxide. *Physiol. Behav.* **1990**, *47*, 1089–1098. [CrossRef]
53. Isonaka, R.; Gibbons, C.H.; Wang, N.; Freeman, R.; Goldstein, D.S. Association of innervation-adjusted alpha-synuclein in arrector pili muscles with cardiac noradrenergic deficiency in autonomic synucleinopathies. *Clin. Auton. Res.* **2019**, *29*, 587–593. [CrossRef]
54. Kim, J.S.; Park, H.E.; Oh, Y.S.; Song, I.U.; Yang, D.W.; Park, J.W.; Lee, K.S. (123)I-MIBG myocardial scintigraphy and neurocirculatory abnormalities in patients with dementia with Lewy bodies and Alzheimer's disease. *J. Neurol. Sci.* **2015**, *357*, 173–177. [CrossRef]
55. Lamotte, G.; Holmes, C.; Sullivan, P.; Lenka, A.; Goldstein, D.S. Cardioselective peripheral noradrenergic deficiency in Lewy body synucleinopathies. *Ann. Clin. Transl. Neurol.* **2020**, *7*, 2450–2460. [CrossRef]

56. Goldstein, D.S.; Holmes, C.; Sullivan, P.; Lopez, G.; Gelsomino, J.; Moore, S.; Isonaka, R.; Wu, T.; Sharabi, Y. Cardiac noradrenergic deficiency revealed by 18F-dopamine positron emission tomography identifies preclinical central Lewy body diseases. *J. Clin. Investig.* **2024**, *134*, e172460. [CrossRef]
57. Goldstein, D.S.; Pekker, M.J.; Eisenhofer, G.; Sharabi, Y. Computational modeling reveals multiple abnormalities of myocardial noradrenergic function in Lewy body diseases. *JCI Insight* **2019**, *5*, e130441. [CrossRef]
58. Goldstein, D.S.; Kopin, I.J.; Sharabi, Y. Catecholamine autotoxicity. Implications for pharmacology and therapeutics of Parkinson disease and related disorders. *Pharmacol. Ther.* **2014**, *144*, 268–282. [CrossRef]
59. Goldstein, D.S.; Pekker, M.J.; Sullivan, P.; Isonaka, R.; Sharabi, Y. Modeling the Progression of Cardiac Catecholamine Deficiency in Lewy Body Diseases. *J. Am. Heart Assoc.* **2022**, *11*, e024411. [CrossRef]
60. Troshev, D.; Bannikova, A.; Blokhin, V.; Pavlova, E.; Kolacheva, A.; Ugrumov, M. Compensatory Processes in Striatal Neurons Expressing the Tyrosine Hydroxylase Gene in Transgenic Mice in a Model of Parkinson's Disease. *Int. J. Mol. Sci.* **2023**, *24*, 16245. [CrossRef]
61. Rao, R.; Androulakis, I.P. Allostatic adaptation and personalized physiological trade-offs in the circadian regulation of the HPA axis: A mathematical modeling approach. *Sci. Rep.* **2019**, *9*, 11212. [CrossRef]

Journal of
*Personalized
Medicine*

MDPI

Article

Neuropsychiatric Manifestations of Mast Cell Activation Syndrome and Response to Mast-Cell-Directed Treatment: A Case Series

Leonard B. Weinstock [1,*], Renee M. Nelson [2] and Svetlana Blitshteyn [2,3]

1 Independent Researcher, Specialists in Gastroenterology, St. Louis, MO 63141, USA
2 Department of Neurology, Jacobs School of Medicine and Biomedical Sciences, University at Buffalo, Buffalo, NY 14203, USA; reneenel@buffalo.edu (R.M.N.); sb25@buffalo.edu (S.B.)
3 Dysautonomia Clinic, Williamsville, NY 14221, USA
* Correspondence: lw@gidoctor.net

Abstract: Mast cell activation syndrome (MCAS) is an immune disease with an estimated prevalence of 17%. Mast cell chemical mediators lead to heterogeneous multisystemic inflammatory and allergic manifestations. This syndrome is associated with various neurologic and psychiatric disorders, including headache, dysautonomia, depression, generalized anxiety disorder, and many others. Although MCAS is common, it is rarely recognized, and thus, patients can suffer for decades. The syndrome is caused by aberrant mast cell reactivity due to the mutation of the controller gene. A case series is presented herein including eight patients with significant neuropsychiatric disorders that were often refractory to standard medical therapeutics. Five patients had depression, five had generalized anxiety disorder, and four had panic disorder. Other psychiatric disorders included attention-deficit hyperactivity disorder, obsessive compulsive disorder, phobias, and bipolar disorder. All eight patients were subsequently diagnosed with mast cell activation syndrome; six had comorbid autonomic disorders, the most common being postural orthostatic tachycardia syndrome; and four had hypermobile Ehlers-Danlos syndrome. All patients experienced significant improvements regarding neuropsychiatric and multisystemic symptoms after mast-cell-directed therapy. In neuropsychiatric patients who have systemic symptoms and syndromes, it is important to consider the presence of an underlying or comorbid MCAS.

Keywords: anxiety; depression; dysautonomia; mast cell activation syndrome; panic disorder; POTS

1. Introduction

Mast cell activation syndrome (MCAS) presents with heterogenous multisystemic inflammatory and allergic manifestations [1–3]. MCAS is characterized by patterns of aberrant mast cell (MC) overactivity [2]. Mast cell activation disease (MCAD), which includes MCAS and mastocytosis, is associated with neuropsychiatric disorders, including various types of dysautonomia, neuropathy (including small fiber neuropathy), myalgia, migraine, headache, cognitive dysfunction, restless legs syndrome, sleep disturbance, non-pulsatile tinnitus, depression, generalized anxiety, and panic attacks [2,4]. MCAS is the most common variant of MCAD and has an estimated prevalence of 17% in the general population [5]. Despite a significant prevalence, this hyperactive immune disorder is usually not considered in the differential diagnosis in patients with multisystemic symptoms [1,6]. This is in part due to its relatively recent discovery (2007) and it is generally not included in medical school curriculum [7].

The heterogeneity of MCAS is vast, with symptoms and syndromes across various domains including constitutional, dermatologic, ophthalmologic, otologic, oropharyngeal, lymphatic, pulmonary, cardiovascular, gastrointestinal, genitourinary, musculoskeletal, neurologic, psychiatric, metabolic, hematologic, and immunologic systems (Table 1) [2].

Patients with unrecognized, undiagnosed MCAS will often see multiple specialists and ultimately stop reporting symptoms owing to poor experiences with the medical system. Unfortunately, these patients are at risk of being misdiagnosed with somatization disorder or Munchausen's syndrome.

Table 1. Common symptoms of mast cell activation syndrome. There are many heterogenous phenotypes that vary according to mast cell location, number, and ability to degranulate specific mediators. Symptoms may be continuous or intermittent and be of various levels of severity.

Constitutional	Fatigue, subjective hyperthermia and/or hypothermia, sweats, change in appetite, weight gain/loss, chemical/physical sensitivities, poor healing
Dermatologic	Urticaria, itch, flushing, hemangiomas with itch/pain, various rashes, telangiectasias, striae, skin tags, folliculitis, ulcers, eczema, angioedema, alopecia, onychodystrophy
Ophthalmologic	Irritated, "dry" eyes, difficulty focusing, blepharospasm
Otologic	Tinnitus, hearing loss, coryza, rhinitis, nasal congestion, epistaxis
Oropharyngeal	Pain, burning, leukoplakia, ulcers, angioedema, dysgeusia, dental and/or periodontal inflammation/decay
Lymphatic	Lymphadenopathy, rare splenomegaly
Pulmonary	Dry cough, dyspnea (difficulty taking a deep breath), wheezing, obstructive sleep apnea
Cardiovascular	Presyncope, hypertension, blood pressure lability, palpitations, edema, chest pain, allergic angina (Kounis syndrome)
Gastrointestinal	Dyspepsia, gastroesophageal reflux, abdominal pain, nausea, vomiting, diarrhea and/or constipation, gastroparesis, angioedema, dysphagia (usually proximal), bloating (post-prandial or spontaneous), malabsorption
Genitourinary	Menorrhagia, pelvic pain, endometriosis, vulvodynia, vaginitis, dysmenorrhea, miscarriages, infertility, dysuria
Musculoskeletal	Myalgias, migratory bone/joint pain, osteopenia/osteoporosis
Neurologic	Headache, migraine, sensory neuropathies, dysautonomia, episodic weakness, seizure disorders, non-epileptic seizures, cognitive dysfunction, insomnia, hypersomnolence, restless leg syndrome
Psychiatric	Depression, anger/irritability, mood lability, anxiety, panic, obsession–compulsion, attention deficit/hyperactivity
Hematologic	Easy bruising, polycythemia, anemia
Immunologic	Hypersensitivity reactions, increased risk for malignancy and autoimmunity, impaired healing, increased susceptibility to infection

MCs can be located adjacent to blood vessels along the blood–brain barrier (BBB) and interact with microglia, astrocytes, and blood vessels through stored or synthesized neuroactive mediators [8]. As the effector cells of the innate immune system, MCs are first to respond to injury, releasing proinflammatory signals to which microglia respond through the production of cytokines, chemokines, glutamate, and reactive oxygen species [9]. Mature MCs can migrate to the brain from the vascular system, and evidence suggests that MCs themselves may disrupt the integrity of the BBB [8]. Signaling between MCs and microglia modulates immunologic responses to inflammation, infection, trauma, and stress. In the setting of prolonged neuroinflammation, these controls may be less effective, and aberrant inflammatory responses may ensue.

The severity of symptoms ranges from mild to life-threatening when anaphylaxis is present. The degree of morbidity is related to the quantity of the affected mast cells (MCs), the number of mutations within the mast cell (MC) lineage, the specific organ involvement, the presence of comorbid postural orthostatic tachycardia syndrome (POTS) and hypermobile Ehlers-Danlos syndrome (hEDS), and the impact of triggers [10–13]. There are many MC triggers, including diet, stress, estrogens, excipients, and a variety of infections [14,15]. The long-lasting, often hidden triggers on which treatment can have a significant impact include small intestinal bacterial overgrowth, mycotoxin and chemical exposures, and heavy metal toxicity [16–19]. We present a case series of patients diagnosed with refractory neuropsychiatric disorders who were subsequently diagnosed with MCAS and whose conditions improved with MC-directed treatment.

2. Materials and Methods

This is a retrospective case series of eight patients who experienced chronic neuropsychiatric symptoms and disorders and were subsequently diagnosed by the authors with MCAS. We observed that these patients experienced significant improvement regarding their psychiatric symptoms when their MCAS was treated with MC-directed therapy. Criteria for inclusion in this case series were a diagnosis of a psychiatric disorder by a primary care physician and/or a psychiatrist and a diagnosis of MCAS (Table 2) [6]. Exclusion criteria included mastocytosis or having symptoms and signs best explained by a medical condition other than MCAS. Comorbid syndromes including autonomic disorders, such as postural orthostatic tachycardia syndrome (POTS), and joint hypermobility syndromes, such as hypermobile Ehlers-Danlos syndrome (hEDS), were assessed. All subjects gave their informed consent for inclusion before they participated in the study. The study was conducted in accordance with the Declaration of Helsinki. The study was approved by the University at Buffalo Jacobs School of Medicine Institutional Review Board (STUDY00006936).

Table 2. Diagnostic criteria for mast cell activation syndrome.

Consensus 1:
1. Severe, recurrent mast cell symptoms, which often include anaphylaxis and involve 2 or more organs, including urticaria, flushing, pruritus, angioedema, nasal congestion/pruritus, wheezing, throat swelling, hoarseness, headache, hypotensive syncope, tachycardia, cramping, and diarrhea; 2. Increased mast cell mediators "preferably tryptase or increased tryptase from baseline plus 20% + 2 during an attack" or "less specific mediators" (plasma histamine or prostaglandin D2, serum heparin, urine N-methylhistamine); 3. Response to mast-cell-directed therapy.
Consensus 2:
Presence of 2 systems with typical mast cell activation symptoms and ≥1 of the following:
1. Positive mast cell mediators (plasma histamine or prostaglandin D2, serum heparin, tryptase, and chromogranin A, or urine N-methylhistamine, leukotriene E4, and 2, 3 dinor prostaglandin F2 alpha); 2. Biopsy showing >20 mast cells per high-power field;
Positive clinical response to mast-cell-directed therapy

Although many FDA-approved medicines have been studied and used for MCAS, there is no FDA-approved protocol for MCAS [14]. As part of our standard approach, we advise a 3-week trial on a gluten-free, dairy-free, and low-histamine diet. Medical therapy starts with a combination of non-sedating histamine receptor 1 and histamine receptor 2 blockers twice daily [9]. If the response is inadequate or the patients have significant symptoms, they receive additional over-the-counter MC stabilizing agents such as vitamins C and D and quercetin, a flavonoid. For those with extensive symptoms, they are administered the Step 1 MC-directed therapy, which includes antihistamines, vitamins C and D, quercetin, and the addition of low-dose naltrexone (LDN). The use of LDN has been reported to be effective in treating MCAS and depression [20,21]. The administration of additional medicines including immune modulators and chemotherapy are used for refractory MCAS [14,22].

3. Results

The subjects included seven females and one male with a mean age of 36 years (ages ranged from 18 to 71 years). Prior neuropsychiatric diagnoses, medical therapy, past medical history, new diagnoses, and the outcomes of MC-directed therapy are outlined in Table 3. Five had depression and one had bipolar disease. Two of these patients had attempted suicide as teenagers. Five had generalized anxiety disorder and four had panic disorder. Seven subjects had additional disorders: attention-deficit/hyperactivity disorder, obsessive–compulsive disorders, phobias, Tourette's syndrome, and narcolepsy.

Table 3. Neuropsychiatric manifestations of mast cell activation syndrome patients and response to mast-cell-directed therapy: cases 1–4. Neuropsychiatric manifestations of mast cell activation syndrome and response to mast-cell-directed treatment: cases 5–8.

N	1	2	3	4
Age (years), sex	47, female	50, female	37, female	71, female
Prior psychiatric diagnoses	GAD, OCD, phobia	GAD, panic disorder	Bipolar disorder (suicide attempt age 15), GAD, ADHD, Tourette's, narcolepsy	MDD (suicide attempt age 16)
Clinical course in childhood and adolescence	Anaphylaxis to nuts and antibiotics	None	Brain fog, diarrhea, urticaria, self-abusive behavior, asthma	Headaches, recurrent viral infections, hives, edema with insect bites, allergies, nausea, abdominal pain, depression, menorrhagia
Clinical course in adulthood	Postpartum phobias, rashes, facial swelling, pruritus, syncope, tachycardia, migraine	Syncope/presyncope during pregnancy, pacemaker for bradycardia, tachycardia, blurred vision, anxiety, joint pain	Depression (daily suicidal ideation), mania, hallucinations, anxiety, fatigue, abdominal pain, nausea, myalgia, hives, bone pain, episodic hypertension, bedridden 4 days/week	Depression (daily suicidal ideation), pelvic pain leading to hysterectomy age 21, tinnitus, chest and body pain, interstitial cystitis
Prior psychiatric therapy	Multiple SSRIs without efficacy	Prescribed SSRI: elected not to take it	1 SSRI, 2 SSRNIs, 2 anti-psychotics, 3 benzos, lamotrigine, atomoxetine, dextroamphetamine, guanfacine	3 classes of anti-depressants—multiple agents, lithium, and ECT
New diagnoses	MCAS, hEDS, NCS, IST	MCAS, NCS, labile hypertension	MCAS, POTS, RLS, labile hypertension	MCAS, POTS
Mast cell treatment	Hydroxyzine, cetirizine daily. Prednisone PRN flares	Cetirizine and famotidine daily.	Step 1 therapy, LDN. Maintained on aripiprazole, dextro-amphetamine, and lamotrigine	Antihistamines 1 and 2, hydroxyurea
Outcomes of mast cell treatment on neuropsychiatric conditions	Complete response: works full time	Complete response: works full time. Tachycardia, syncope, flushing, and anxiety resolved	Partial response: works part time	Complete response: independent in ADLs and iADLs. Depression resolved

Table 3. *Cont.*

N	5	6	7	8
Age (years), sex	18, male	18, female	19, female	33, female
Prior neuropsychiatric diagnoses	Panic disorder, GAD, MDD	Panic disorder, GAD, MDD	Panic disorder, MDD	Panic disorder, GAD, MDD
Clinical course in childhood and adolescence	Brain fog, fatigue, rhinitis, diarrhea, abdominal pain with gluten	Constipation, diarrhea, dysphagia, heartburn, nausea, eczema, headache, menorrhagia, syncope	Nausea, diarrhea, menorrhagia, flushing, fatigue, brain fog, tinnitus	Headache, multiple viral infections
Clinical course in adulthood	Myalgias	Constipation, diarrhea, dysphagia, heartburn, nausea, eczema, headache, menorrhagia, syncope	Weight loss, nausea, diarrhea, menorrhagia, flushing, fatigue, brain fog, tinnitus	Nausea, pain, fatigue, weakness, tinnitus, palpitations, flushing, presyncope, migraine, brain fog, hives, itch, bone pain
Prior psychiatric therapy	Escitalopram	Escitalopram, buspirone	Desvenlafaxine, fluvoxamine, fluoxetine	None
New diagnoses	MCAS	MCAS, RLS, hEDS	MCAS, POTS, hEDS	MCAS, POTS, hEDS
Mast cell treatment	Step 1, LDN	H1/2 blockers, LDN, buspirone PRN anxiety	Step 1, LDN	GFD, Step 1, LDN
Outcome on mast cell treatment for neuropsychiatric conditions	Complete response: Able to return to college after withdrawal	Complete response: Able to attend college after home schooling	Marked improvement: Able to return to college. Regained 15 pounds	Complete response: Able to work full time

Abbreviations: ADHD, attention-deficit/hyperactivity disorder; benzos, benzodiazepines; ECT, electroconvulsive therapy; GDF, gluten-free diet; GAD, generalized anxiety disorder; hEDS, hypermobile Ehlers-Danlos syndrome; IST, inappropriate sinus tachycardia; LDN, low-dose naltrexone; MCAS, mast cell activation syndrome; MDD, major depressive disorder; NCS, neurocardiogenic syncope; OCD, obsessive–compulsive disorder; POTS, postural orthostatic tachycardia syndrome; PRN, as needed; SSRI, selective serotonin reuptake inhibitor. Note: Step 1 therapy is a combination mast-cell-directed oral therapy using various over-the-counter histamine 1 receptor antagonists (variety of brands, twice daily), histamine 2 receptor antagonist (famotidine 20 mg twice daily), quercetin (1000 mg twice a day), sustained release vitamin C (1000 mg daily), and vitamin D (2000–5000 mg daily depending on vitamin D level).

We subsequently diagnosed significant disorders: MCAS in eight subjects, POTS in four, labile blood pressure in two, inappropriate sinus tachycardia in one, neurocardiogenic syncope in one, migraine in two, and restless legs syndrome in two. Four subjects had hEDS. All subjects' symptoms improved with MC-directed therapy, as shown in Table 3. Six subjects experienced significant improvement regarding their psychiatric disorder so they could return to work, high school, or college. One woman could now function well as a homemaker. Her daily suicidal ideation, which had persisted for decades, also ceased. One subject improved enough to work part time.

3.1. Illustrative Case: Patient 1

The patient was a 47-year-old physician who was healthy except for episodes of anaphylaxis in response to nuts in her childhood and adolescence. In her twenties, she experienced symptoms of weight loss, lymphadenopathy, night sweats, and fatigue, and at one point, she was thought to have lymphoma, which was subsequently ruled out. She was treated with amoxicillin for presumed bacterial infection, which caused a rash. At that time, her treating physicians thought she may have had mononucleosis. Subsequently, she experienced two bouts of shingles. She experienced no psychiatric symptoms and received no psychiatric diagnosis in childhood, adolescence, or in her twenties and felt that she was healthy and athletic prior to her first pregnancy. During her first pregnancy at age 35, she developed severe nausea associated with weight loss and was diagnosed with hyperemesis gravidarum. During her first pregnancy, she was treated with IV hydration via a central venous line on a temporary basis with good results, and the central venous line was removed toward the end of her pregnancy. She subsequently had uneventful labor, vaginal delivery, and postpartum period and was able to return to work full time without symptoms. She developed a severe phobia of blood 1.5 years post-partum, along with generalized anxiety and obsessive symptoms. Her second pregnancy was at 38 years of age, which was again associated with nausea and food intolerances, but she was able to maintain her weight and did not require a central venous line. However, she did receive intravenous hydration every few weeks through a peripheral line for symptoms thought to be due to dehydration. After an uneventful delivery, her phobia of blood worsened, and she developed compulsive rituals as well as fatigue. A few years later, she developed recurrent rashes, periorbital swelling, flushing episodes (Figure 1), rapid heart rate upon standing or with minimal exertion, and chronic constipation. She also developed presyncope associated with diarrhea and tachycardia, which were often triggered by taking a shower. These symptoms persisted for the next 6 years without an explanation or an identifiable etiology, despite receiving evaluations from a variety of specialists. Due to a severe fear of the sight of blood, anxiety, and obsessive–compulsive disorder, which was diagnosed by a primary care physician, she had to stop working and became housebound after the second pregnancy. She denied depressive symptoms or suicidal ideations. She did report Raynaud's phenomenon as well as recurrent lower-back pain for at least 15 years, which was triggered by prolonged sitting. She was diagnosed with hypermobility spectrum disorder, though she denied chronic muscle or joint pain. She did experience easy bruising and bruxism. Prior to her second pregnancy, she used to run for at least 30 min several times per week. After her second pregnancy, she developed significant exercise intolerance due to resting and exertional tachycardia. She described having a heart rate of 150 beats per minute after walking for only 5 min. Sertraline 200 mg was initiated for OCD and anxiety, which resulted in a partial improvement in psychiatric symptoms; as such she was well enough to briefly leave the house. She reported experiencing chronic insomnia for many years, sleeping only 3 to 5 h per night. Clonazepam was prescribed at a dose of 0.5 mg to 1 mg at bedtime, but it did not prolong her sleep duration. She experienced frequent flushing and angioedema, sometimes triggered by stress, but most of the time, the trigger was unknown. She could not identify any potential food triggers. A gluten-free, low-histamine diet failed to help. After receiving an intramuscular cortisone injection for the treatment of a rash, she felt significantly better for several weeks, both regarding physical and psychological symptom improvement. At the time of her presentation, her most disabling symptoms were obsessive thoughts regarding her fear of seeing blood, fatigue, exercise intolerance, resting and postural tachycardia, and recurrent pruritic facial and neck rash.

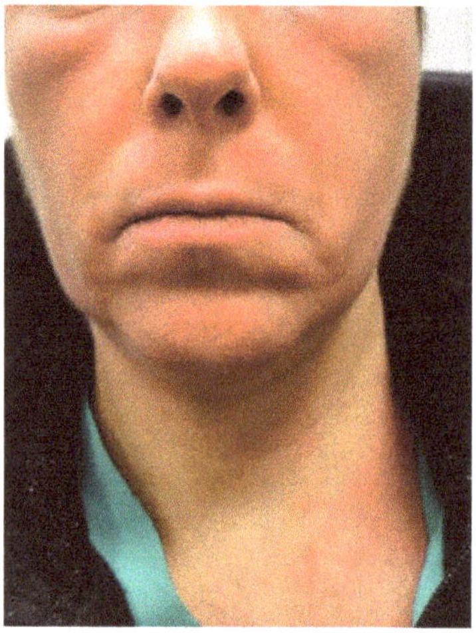

Figure 1. Photo of the periorbital edema with facial and neck flushing provided by patient 1.

A tilt table test demonstrated inappropriate sinus tachycardia (IST) and neurocardiogenic syncope after a nitroglycerin challenge. Diagnostic tests were also remarkable, showing low serum ferritin level, elevated platelets, and mildly reduced IgG1 subclass. Serum and urine MC mediators, including serum tryptase, serum histamine, serum and urine chromogranin, and prostaglandins, were in the normal range at baseline testing, during which she was not experiencing a flare.

Given the clinical features of allergic symptoms and excellent response to antihistamines and steroids, a clinical diagnosis of MCAS was made by an allergist based on the Consensus 2 criteria [6]. Daily hydroxyzine at a dose of 25 mg was initiated and increased to twice a day, along with 10 mg cetirizine daily. The patient reported significant improvement and near complete resolution of both her phobia of blood and obsessive–compulsive thoughts and rituals. Additionally, her elevated resting and exertional heart rate decreased after the implementation of antihistamines without the use of heart-rate-controlling medications typically used for the treatment of IST and other autonomic disorders, such as beta blockers or an I-channel blocker. Given her physical and psychological improvement, she was able to return to work in healthcare full time and resume an exercise training program.

3.2. Illustrative Case: Patient 4—Personal Account

"Here is my story of a lifelong battle with depression, mood swings and healing after a diagnosis and treatment of MCAS. I had depressive symptoms as a little girl, which escalated after I was molested at the age of 11. By the time I was 16, I tried to kill myself. In my early twenties I escalated to severe mood swings going into mania for 6 to 7 days at a time never sleeping or even lying down to an inability to stay awake for days on end. I would go into a rage at the drop of a hat and at times would lock myself in the bathroom to keep away from my sons so they would not receive the outcome of my rage. There was even a time I thought I should give my boys up for adoption because I believed they deserved a better mother than I was able to be. Along with the mood swings I also developed severe panic attacks".

"I saw a psychiatrist who diagnosed me with bipolar disorder. I then started taking medications which would help a little for a short time, but then I'd be right back where I started. I was eventually put on Lithium that helped the symptoms, but I developed severe swelling in my abdomen to the point I had to wear maternity clothes and vomited around the clock. My mood and desire to die was also related to severe pain from a torn lumbar disc that went undiagnosed for 22 years. Between the pain and depression, I continued to get worse, until I attempted suicide again in 1990. I continued to try different medications but had drug reactions and a never-ending circle with depression and thoughts of suicide wrapped in the middle".

"At age 55 I was diagnosed with breast cancer and underwent a mastectomy and chemotherapy. Shortly after chemo I started having multiple symptoms, including heart palpitations up to 200 bpm just walking across the room as well as chest pain and dangerously low blood pressure, I was finally diagnosed with POTS. I also had esophageal spasms, interstitial cystitis, migraine headache and episodes of severe vomiting. I also started suffering severe shortness of breath and was diagnosed with asthma and vocal cord dysfunction. I kept asking for help from doctors but was told it was all in my head and I just needed to see a psychiatrist! I knew all of this was not in my head, but when you have a history of depression and bipolar disorder, I found that very few doctors take you seriously".

"My primary care physician finally referred me to a neurologist for POTS, and my life changed at that appointment when he diagnosed me with MCAS. He assured me my symptoms were not in my head but in my brain. He started me on the MCAS protocol, and although some symptoms were better the depression was not. He then referred me to Dr. Weinstock who tried other medications and eventually added a chemotherapy drug called hydroxyurea. Shortly after starting hydroxyurea, my suicidal thoughts finally stopped, and I have not had any desire to die since. I am truly happy for the first time in my life".

4. Discussion

We present a case series of eight patients with refractory neuropsychiatric disorders who experienced significant improvement after subsequent diagnosis and treatment of MCAS. Previously, these patients exhibited either no response or poor response to a variety of psychiatric medications and psychotherapies. Electroshock therapy had been used in the eldest patient in the cohort. These patients often presented to psychiatrists in their early teenage years, but some developed neuropsychiatric symptoms as an adult, with pregnancy or the postpartum period being a precipitating or exacerbating event. The female predominance in our case series was similar to other MCAS studies where the female to male ratio was over 80% [23,24]. Most of the patients in our case series responded to simple Step 1 MC-directed therapy. The eldest of our patients was a 71-year-old woman who had experienced severe, lifelong depression. Low-dose hydroxyurea (500 mg per day) was added to her regimen to treat refractory gastrointestinal symptoms. To her surprise, her daily suicide ideation resolved for the first time in four decades. Hydroxyurea has been used successfully to treat general systemic symptoms in both refractory MCAS and mastocytosis [22]. This drug is an oral ribonucleotide reductase inhibitor which is used at a high dosage in the treatment of chronic myeloproliferative neoplasms and at a low dosage in sickle cell anemia, where there is evidence that MC activation causes increased cytokines and joint pain [25].

5. Mast Cells

We theorize that MCAS-associated neuropsychiatric disorders could be caused by abnormal MCs in the central and/or peripheral nervous system or indirectly by circulating MC mediators that lead to inflammation in the nervous system. MCs, known as immune and pro-inflammatory effector cells, are present in the meninges and are implicated in the pathophysiology of migraine via neuropeptide release, vasodilation, and plasma and protein extravasation, which can lead to MC degranulation. Since MCs release hundreds of various mediators, including histamines, tryptases, and leukotrienes, the degranulation of meningeal MCs contributes to the sensitization of trigeminal vascular afferent processing. This MC-mediated pathway is thought to be one of the mechanisms underlying migraine pain pathophysiology, and migraine is one of the most common comorbidities noted in patients with MCAS [26].

Similar to our case series, patients with MCAS can have comorbid autonomic dysfunction. While the mechanisms have not been explored in detail, a recent study linked the parasympathetic nervous system and MCs via its findings, which suggest that the endogenous acetylcholine activates the meningeal MCs [27]. Further studies are needed to delineate the complex interplay between the autonomic nervous system, MCs, and the connective tissues of the meninges, cerebral vasculature, and other structures important to the pathophysiology of the triad of dysautonomia, MCAS, and hypermobility spectrum disorders often observed in clinical practice [28].

Additionally, circulating autoantibodies could affect the brain and autonomic nervous system due to an MC-induced hyperpermeable BBB and/or an abnormally functioning blood–cerebrospinal fluid barrier [8,29]. The role of MC activation in a variety of neuropsychiatric disorders has been studied in humans and in animal models [30–37]. Magnetic resonance imaging has demonstrated morphological and functional abnormalities in the brains of mastocytosis patients with neuropsychiatric complaints [38]. Using the same technique, a MCAS patient with depression also exhibited the same radiographic finding [39]. In a case series of 139 patients with mastocytosis, 49% had depression [31]. In another series of 288 mastocytosis patients, the prevalence of depression was 64%, with 56% described as having moderate and 8% severe depression [30].

6. Histamine and Histamine Receptors

Histamine, a major MC mediator, is a known neurotransmitter in the central nervous system. Histamine is a monoamine that is metabolized from the precursor histidine and is released into some of the neuronal synapses, as well as into the blood stream, where it acts as a hormone. Histamine is also a known neuromodulator since it regulates the release of other neurotransmitters, such as acetylcholine, norepinephrine, and serotonin [40]. The histamine receptors H1, H2, H3, and H4 are a class of G-protein–coupled receptors which bind to histamine as their primary endogenous ligand [41,42]. The H1 receptor mediates immediate hypersensitive reactions, such as wheezing, itching, coughing, and hypotension; the H2 receptor affects gastric mucosa, vascular smooth muscle, fat cells, basophils, and neutrophils and inhibits antibody synthesis, T-cell proliferation, and cytokine production; the H3 receptor decreases the release of acetylcholine, serotonin, and norepinephrine neurotransmitters in the central nervous system; and the H4 receptor is implicated in mast cell chemotaxis and regulating immune responses [42–44].

Histamine is known to contribute to the regulation of sleep and wakefulness, and histamine blocking is a well-known pharmacological approach used to induce sleep. Low levels of histamine have been shown to correlate with schizophrenia, and an altered histaminergic system has been found in the nigrostriatal network in Parkinson's disease [40]. Postmortem studies have revealed alterations in the histaminergic system in neurological and psychiatric diseases. Brain histamine levels are decreased in Alzheimer's disease patients, whereas abnormally high histamine concentrations are found in the brains of Parkinson's disease and schizophrenic patients [40]. Low histamine levels are associated with convulsions and seizures [40]. The release of histamine is altered in response to different types of brain injury; for example, the increased release of histamine in an ischemic brain trauma might play a role in recovery following neuronal damage [43]. Neuronal histamine is also involved in pain, and drugs that increase brain and spinal histamine concentrations have antinociceptive properties. Histaminergic drugs, most importantly histamine H3 receptor ligands, have shown efficacy in many animal models of the neurologic disorders, and clinical trials to determine the efficacy and safety of these drugs in humans are needed [43].

Histamine has a significant underexplored potential to provide targets for many CNS disorders. The histamine system has been suggested as a possible target for the treatment of psychiatric disorders, and drugs that modulate this system have been proposed as cognitive enhancers [44]. Greater understanding of histamine, histamine receptors, and histaminergic pathways in the central and peripheral nervous systems is particularly relevant for the development of novel pharmacological treatments for neurologic and psychiatric disorders [42,45–47].

Lastly, histamine may cause the increased permeability of the blood–brain barrier. It also significantly influences neuroendocrine control, including the behavioral state, biological rhythms, energy metabolism, thermoregulation, fluid balance, stress, and reproduction. In addition to being a neurotransmitter and neuromodulator, histamine is also associated with the functioning of the immune system. During an immune reaction, histamine is released and contributes to the physiologic changes necessary for the immune system to fight a pathogen, including an increase in blood pressure, temperature, swelling, and bronchial constriction [8,42,45,47].

7. Mast-Cell-Directed Treatment

Standard psychiatric medicines are frequently prescribed for patients presenting with depression and anxiety; however, a significant subset of patients is refractory to these treatments or experiences adverse events. Our case series suggests that when MCAS is suspected and then diagnosed, MC-directed therapy can be effective in improving neuropsychiatric manifestations. Treatment with antihistamines, MC-stabilizing agents, and other pharmacologic modalities such as LDN, along with non-pharmacologic approaches including avoiding symptomatic triggers and adopting low-histamine and gluten/dairy-free diets, can be effective, are inexpensive, and have a low side effect profile compared to standard antidepressant and antianxiety therapies. Another consideration regarding the intolerance to standard psychiatric medications experienced by the general population, particularly for MCAS patients, is the frequency of reaction to excipients [48].

Benzodiazepines are used in the treatment of anxiety and panic attacks. These medications have been demonstrated to have an inhibitory action on the pro-inflammatory effector functions of MCs [49]. In addition, there is evidence in the literature that supports the role of histamine as a neurotransmitter in stress-related disorders [50]. Microglia express all four histamine receptors, with selective upregulation of H1R and H4R [51]. Astrocytes express the H1R and H2R histamine receptors [52]. This may be another mechanism for the effect of histamine receptor blockers in MCAS and psychiatric disorders. Nevertheless, due to the adverse effects and addictive potential associated with chronic benzodiazepine use, benzodiazepines should not be routinely prescribed and should be reserved for patients who are refractory to other non-addictive MC therapies.

For severe cases of MCAS, immune modulators can be helpful. These medications include hydroxyurea and various tyrosine kinase inhibitors (TKI). Masitinib, a TKI, has been used as effective treatment for drug-refractory depression in mastocytosis and is currently being studied in MCAS patients [30,53] [Clintrials.gov NCT05449444]. In the largest mastocytosis case series to date, of 288 patients treated with masitinib, 67% experienced a significant improvement with regard to depression overall, and 75% recovered [30]. In a case report, a MCAS patient with severe MCAS and postural orthostatic tachycardia syndrome experienced significant improvement with regard to depression, anxiety, and dysautonomia symptoms using intravenous immunoglobulin, LDN, and antibiotic treatment for small intestinal bacterial overgrowth [21].

8. Autonomic Dysfunction and MCAS

Autonomic dysfunction is common in MCAS patients, possibly due to the MC mediator effects on the central autonomic networks in the brain, peripheral autonomic and small nerve fibers, and the vasodilatory effects of the mediators on blood vessels and via other yet unidentified mechanisms. One study identified clinical evidence of MC hyperactivity in 64% of their patients with POTS, 66% of whom received at least one positive laboratory finding suggestive of MC hyperactivity [28]. In another study, the percentage of MCAS diagnosis within a group of POTS and hEDS patients was 31% in comparison with 2% in a group without POTS or hEDS [54]. Small fiber neuropathy and cerebral hypoperfusion may share pathophysiologic mechanisms in MCAS, dysautonomia, and hEDS [55]. While the true prevalence of MCAS in patients with POTS or hEDS is unknown, considering the lack of awareness of MCAS as a diagnostic entity among clinicians and the difficulty of confirming diagnosis objectively, most of the patients in our series had comorbid autonomic disorders, with POTS being the most common diagnosis.

9. Limitations

The limitations of this study include those inherent to the nature of a retrospective chart review, the small sample size, the subjectivity of patient-reported functional improvement following treatment, the lack of a control group, the referral bias and heterogeneity of various mast-cell-directed treatment, the influence of psychiatric medications, and the treatment approaches for autonomic comorbidities and other comorbidities. In addition, we recognize that there is some controversy regarding the diagnostic criteria for MCAS [6]. There are limitations related to our patient selection and the generalizability of our findings.

10. Clinical Relevance in Personalized Care

Although the literature on MCAS, dysautonomia, and hypermobility spectrum disorders is relatively limited, patients who have all three conditions as a triad are often encountered in clinical practice [56]. Many of these patients have been sick for years or decades, seen multiple physicians of various specialties, tried a wide variety of medications and supportive therapies with limited improvement, and have experienced significant functional impairment. While neurologic, autonomic, and psychiatric comorbidities in these patients are numerous, many patients are misdiagnosed with psychiatric diagnoses such as somatic symptom disorder, medically unexplained symptoms, functional neurologic disorders, somatization disorder, factitious disorder, or malingering. Some parents of children and teens with the triad have been wrongfully accused of Munchausen by proxy. The mislabeling of these patients with psychiatric illness as the cause for a systemic illness often leads to inappropriate or misdirected treatment, iatrogenic adverse events, resentment, mistrust on the part of the patient, doctor shopping, non-compliance, medical care avoidance, and psychological symptoms and trauma caused by their negative experience with the healthcare system. As the illustrative cases demonstrate, significant improvement in and even resolution of decades of neuropsychiatric symptoms are possible when an underlying systemic disorder is identified and therapeutic modalities for the underlying systemic disorder are instituted. Although at this time it is unknown whether a relationship between MCAS, dysautonomia, and hypermobility spectrum disorders is rooted in causation or association, we believe that a patient-centered, comprehensive, and personalized approach to neurologic and psychiatric care is essential to accurate diagnosis, effective treatment, and reducing the symptom burden and disability associated with these multisystemic complex chronic disorders.

11. Conclusions

In patients with neuropsychiatric disorders refractory to standard therapy who have systemic symptoms, underlying MCAS should be considered in the differential diagnosis. This is especially the case if they have comorbid POTS, other types of dysautonomia, and/or hEDS. Prospective randomized controlled trials are needed to determine the prevalence of MCAS in patients with treatment-refractory neuropsychiatric disorders, delineate the neurologic and psychiatric manifestations of MCAS, and assess the response to MCAS-targeted treatment.

Author Contributions: L.B.W. and S.B. conceived of the study. All three physicians (L.B.W., S.B. and R.M.N.) performed the research, analyzed the data, and drafted the manuscript. All authors critically reviewed and contributed to the manuscript. All authors have read and agreed to the published version of the manuscript.

Funding: This research received no external funding.

Institutional Review Board Statement: The study was conducted in accordance with the Declaration of Helsinki. The study was approved by the Buffalo Jacobs School of Medicine's Institutional Review Board (STUDY00006936).

Informed Consent Statement: All subjects gave their informed consent for inclusion before they participated in the study.

Data Availability Statement: All of the patient's histories are maintained in the electronic medical records of Blitshteyn and Weinstock.

Conflicts of Interest: The authors declare no conflict of interest.

Abbreviations

BBB: blood–brain barrier; hEDS, hypermobile Ehlers-Danlos syndrome; MC, mast cell; MCs; mast cells; MCAD, mast cell activation disease; MCAS, mast cell activation syndrome; POTS, postural orthostatic tachycardia syndrome.

References

1. Afrin, L.B.; Butterfield, J.H.; Raithel, M.; Molderings, G.J. Often seen, rarely recognized: Mast cell activation disease—A guide to diagnosis and therapeutic options. *Ann. Med.* **2016**, *48*, 190–201. [CrossRef] [PubMed]
2. Afrin, L.B.; Self, S.; Menk, J.; Lazarchick, J. Characterization of Mast Cell Activation Syndrome. *Am. J. Med. Sci.* **2017**, *353*, 207–215. [CrossRef] [PubMed]
3. Hamilton, M.J. Nonclonal Mast Cell Activation Syndrome: A Growing Body of Evidence. *Immunol. Allergy Clin. N. Am.* **2018**, *38*, 469–481. [CrossRef]
4. Afrin, L.B.; Pöhlau, D.; Raithel, M.; Haenisch, B.; Dumoulin, F.L.; Homann, J.; Mauer, U.M.; Harzer, S.; Molderings, G.J. Mast cell activation disease: An underappreciated cause of neurologic and psychiatric symptoms and diseases. *Brain Behav. Immun.* **2015**, *50*, 314–321. [CrossRef] [PubMed]
5. Molderings, G.J.; Haenisch, B.; Bogdanow, M.; Fimmers, R.; Nöthen, M.M. Familial Occurrence of Systemic Mast Cell Activation Disease. *PLoS ONE* **2013**, *8*, e76241. [CrossRef]
6. Afrin, L.B.; Ackerley, M.B.; Bluestein, L.S.; Brewer, J.H.; Brook, J.B.; Buchanan, A.D.; Cuni, J.R.; Davey, W.P.; Dempsey, T.T.; Dorff, S.R.; et al. Diagnosis of mast cell activation syndrome: A global "consensus-2". *Diagnosis* **2020**, *8*, 137–152. [CrossRef]
7. Molderings, G.J.; Kolck, U.W.; Scheurlen, C.; Brüss, M.; Homann, J.; Von Kügelgen, I. Multiple novel alterations in *Kit* tyrosine kinase in patients with gastrointestinally pronounced systemic mast cell activation disorder. *Scand. J. Gastroenterol.* **2007**, *42*, 1045–1053. [CrossRef]
8. Shelestak, J.; Singhal, N.; Frankle, L.; Tomor, R.; Sternbach, S.; McDonough, J.; Freeman, E.; Clements, R. Increased blood-brain barrier hyperpermeability coincides with mast cell activation early under cuprizone administration. *PLoS ONE* **2020**, *15*, e0234001. [CrossRef]
9. Weinstock, L.B.; Pace, L.A.; Rezaie, A.; Afrin, L.B.; Molderings, G.J. Mast Cell Activation Syndrome: A Primer for the Gastroenterologist. *Dig. Dis. Sci.* **2020**, *66*, 965–982. [CrossRef]
10. Molderings, G.J. The genetic basis of mast cell activation disease—Looking through a glass darkly. *Crit. Rev. Oncol.* **2015**, *93*, 75–89. [CrossRef]
11. DiBaise, J.K.; Harris, L.A.; Goodman, B. Postural Tachycardia Syndrome (POTS) and the GI Tract: A Primer for the Gastroenterologist. *Am. J. Gastroenterol.* **2018**, *113*, 1458–1467. [CrossRef] [PubMed]
12. Tinkle, B.; Castori, M.; Berglund, B.; Cohen, H.; Grahame, R.; Kazkaz, H.; Levy, H. Hypermobile Ehlers-Danlos syndrome (a.k.a. Ehlers-Danlos syndrome Type III and Ehlers-Danlos syndrome hypermobility type): Clinical description and natural history. *Am. J. Med. Genet. Part C Semin. Med. Genet.* **2017**, *175*, 48–69. [CrossRef] [PubMed]
13. Häder, T.; Molderings, G.J.; Klawonn, F.; Conrad, R.; Mücke, M.; Sellin, J. Cluster-Analytic Identification of Clinically Meaningful Subtypes in MCAS: The Relevance of Heat and Cold. *Dig. Dis. Sci.* **2023**, *68*, 3400–3412. [CrossRef] [PubMed]
14. Molderings, G.J.; Haenisch, B.; Brettner, S.; Homann, J.; Menzen, M.; Dumoulin, F.L.; Panse, J.; Butterfield, J.; Afrin, L.B. Pharmacological treatment options for mast cell activation disease. *Naunyn-Schmiedeberg's Arch. Pharmacol.* **2016**, *389*, 671–694. [CrossRef]

15. Bernard, Q.; Wang, Z.; Di Nardo, A.; Boulanger, N. Interaction of primary mast cells with Borrelia burgdorferi (sensu stricto): Role in transmission and dissemination in C57BL/6 mice. *Parasites Vectors* **2017**, *10*, 313. [CrossRef]
16. Weinstock, L.B.; Rezaie, A.; Brook, J.B.; Kaleem, Z.; Afrin, L.B.; Molderings, G.J. Small Intestinal Bacterial Overgrowth is Common in Mast Cell Activation Syndrome. *J. Med. Clin. Res. Rev.* **2020**, *4*, 1–13. [CrossRef]
17. Kritas, S.K.; Conti, P. Impact of mold on mast cell-cytokine immune response. *J. Biol. Regul. Homeost. Agents* **2018**, *32*, 763–768.
18. Dearman, R.J.; Betts, C.J.; Humphreys, N.; Flanagan, B.F.; Gilmour, N.J.; Basketter, D.A.; Kimber, I. Chemical Allergy: Considerations for the Practical Application of Cytokine Profiling. *Toxicol. Sci.* **2003**, *71*, 137–145. [CrossRef]
19. Schedle, A.; Samorapoompichit, P.; Füreder, W.; Rausch-Fan, X.H.; Franz, A.; Sperr, W.R.; Slavicek, R.; Simak, S.; Klepetko, W.; Ellinger, E.; et al. Metal ion-induced toxic histamine release from human basophils and mast cells. *J. Biomed. Mater. Res.* **1998**, *39*, 560–567.
20. Mischoulon, D.; Hylek, L.; Yeung, A.S.; Clain, A.J.; Baer, L.; Cusin, C.; Ionescu, D.F.; Alpert, J.E.; Soskin, D.P.; Fava, M. Randomized, proof-of-concept trial of low dose naltrexone for patients with breakthrough symptoms of major depressive disorder on antidepressants. *J. Affect. Disord.* **2017**, *208*, 6–14. [CrossRef]
21. Weinstock, L.B.; Brook, J.B.; Myers, T.L.; Goodman, B. Successful treatment of postural orthostatic tachycardia and mast cell activation syndromes using naltrexone, immunoglobulin and antibiotic treatment. *BMJ Case Rep.* **2018**, *2018*, bcr-2017. [CrossRef]
22. Weinstock, L.B.; Brook, J.B.; Molderings, G.J. Efficacy and toxicity of hydroxyurea in mast cell activation syndrome patients refractory to standard medical therapy: Retrospective case series. *Naunyn-Schmiedeberg's Arch. Pharmacol.* **2022**, *395*, 1441–1447. [CrossRef]
23. Weinstock, L.B.; Brook, J.B.; Brook, J.B.; Blasingame, K.E.; Kaleem, Z.; Afrin, L.B.; Molderings, G.J. Tinnitus in Mast Cell Activation Syndrome: A Prospective Survey of 114 Patients. *J. Otolaryngol. Neurotol. Res.* **2021**, *4*, 92–96.
24. Weinstock, L.B.; Walters, A.S.; Brook, J.B.; Kaleem, Z.; Afrin, L.B.; Molderings, G.J. Restless legs syndrome is associated with mast cell activation syndrome. *Sleep Med.* **2020**, *16*, 401–408. [CrossRef]
25. Vincent, L.; Vang, D.; Nguyen, J.; Gupta, M.; Luk, K.; Ericson, M.E.; Simone, D.A.; Gupta, K. Mast cell activation contributes to sickle cell pathobiology and pain in mice. *Blood* **2013**, *122*, 1853–1862. [CrossRef] [PubMed]
26. Hendriksen, E.; van Bergeijk, D.; Oosting, R.S.; Redegeld, F.A. Mast cells in neuroinflammation and brain disorders. *Neurosci. Biobehav. Rev.* **2017**, *79*, 119–133. [CrossRef]
27. Qian, Y.; Dong, H.; Zhang, X. Mast Cells and Neuroinflammation. *Med. Sci. Monit. Basic Res.* **2014**, *20*, 200–206. [CrossRef]
28. Wang, E.; Ganti, T.; Vaou, E.; Hohler, A. The relationship between mast cell activation syndrome, postural tachycardia syndrome, and Ehlers-Danlos syndrome. *Allergy Asthma Proc.* **2021**, *42*, 243–246. [CrossRef]
29. Gelb, S.; Stock, A.D.; Anzi, S.; Putterman, C.; Ben-Zvi, A. Mechanisms of neuropsychiatric lupus: The relative roles of the blood-cerebrospinal fluid barrier versus blood-brain barrier. *J. Autoimmun.* **2018**, *91*, 34–44. [CrossRef]
30. Moura, D.S.; Sultan, S.; Georgin-Lavialle, S.; Pillet, N.; Montestruc, F.; Gineste, P.; Barete, S.; Damaj, G.; Moussy, A.; Lortholary, O.; et al. Depression in Patients with Mastocytosis: Prevalence, Features and Effects of Masitinib Therapy. *PLoS ONE* **2011**, *6*, e26375. [CrossRef]
31. Georgin-Lavialle, S.; Gaillard, R.; Moura, D.; Hermine, O. Mastocytosis in adulthood and neuropsychiatric disorders. *Transl. Res.* **2016**, *174*, 77–85.e1. [CrossRef] [PubMed]
32. Kempuraj, D.; Selvakumar, G.P.; Thangavel, R.; Ahmed, M.E.; Zaheer, S.; Raikwar, S.P.; Iyer, S.S.; Bhagavan, S.M.; Beladakere-Ramaswamy, S.; Zaheer, A. Mast Cell Activation in Brain Injury, Stress, and Post-traumatic Stress Disorder and Alzheimer's Disease Pathogenesis. *Front. Neurosci.* **2017**, *11*, 703. [CrossRef] [PubMed]
33. Nicoloro-SantaBarbara, J.; Carroll, J.; Lobel, M. Coping, social support, and anxiety in people with mast cell disorders. *Ann. Allergy Asthma Immunol.* **2021**, *127*, 435–440. [CrossRef]
34. SantaBarbara, J.N.; Lobel, M. Depression, psychosocial correlates, and psychosocial resources in individuals with mast cell activation syndrome. *J. Health Psychol.* **2021**, *27*, 2013–2026. [CrossRef] [PubMed]
35. Jendoubi, F.; Severino-Freire, M.; Negretto, M.; Arbus, C.; Paul, C.; Livideanu, C.B. Neuropsychiatric, cognitive and sexual impairment in mastocytosis patients. *Orphanet J. Rare Dis.* **2021**, *16*, 118. [CrossRef]
36. Bidri, M.; Royer, B.; Averlant, G.; Bismuth, G.; Guillosson, J.-J.; Arock, M. Inhibition of mouse mast cell proliferation and proinflammatory mediator release by benzodiazepines. *Immunopharmacology* **1999**, *43*, 75–86. [CrossRef]
37. Ikarashi, Y.; Yuzurihara, M. Experimental anxiety induced by histaminergics in mast cell-deficient and congenitally normal mice. *Pharmacol. Biochem. Behav.* **2002**, *72*, 437–441. [CrossRef]
38. Boddaert, N.; Salvador, A.; Chandesris, M.O.; Lemaître, H.; Grévent, D.; Gauthier, C.; Hermine, O. Neuroimaging evidence of brain abnormalities in mastocytosis. *Transl. Psychiatry* **2017**, *7*, e1197. [CrossRef]
39. Haenisch, B.; Molderings, G. White matter abnormalities are also repeatedly present in patients with systemic mast cell activation syndrome. *Transl. Psychiatry* **2018**, *8*, 95. [CrossRef]
40. Nuutinen, S.; Panula, P. Histamine in neurotransmission and brain diseases. *Adv. Exp. Med. Biol.* **2010**, *709*, 95–107. [CrossRef]
41. Ravhe, I.S.; Krishnan, A.; Manoj, N. Evolutionary history of histamine receptors: Early vertebrate origin and expansion of the H3-H4 subtypes. *Mol. Phylogenet. Evol.* **2020**, *154*, 106989. [CrossRef] [PubMed]
42. Thangam, E.B.; Jemima, E.A.; Singh, H.; Baig, M.S.; Khan, M.; Mathias, C.B.; Church, M.K.; Saluja, R. The Role of Histamine and Histamine Receptors in Mast Cell-Mediated Allergy and Inflammation: The Hunt for New Therapeutic Targets. *Front. Immunol.* **2018**, *9*, 1873. [CrossRef] [PubMed]

43. Tiligada, E.; Kyriakidis, K.; Chazot, P.L.; Passani, M.B. Histamine Pharmacology and New CNS Drug Targets. *CNS Neurosci. Ther.* **2011**, *17*, 620–628. [CrossRef]
44. Lieberman, P.; Hernandez-Trujillo, V.; Lieberman, J.; Frew, A.J. *Clinical Immunology*, 3rd ed.; Rich, R.R., Fleisher, T.A., Shearer, W.T., Schroeder, H.W., Frew, A.J., Weyand, C.M., Eds.; Mosby: St. Louis, MI, USA, 2008; Chapter 89; pp. 1317–1329. ISBN 9780323044042.
45. Cheng, L.; Liu, J.; Chen, Z. The Histaminergic System in Neuropsychiatric Disorders. *Biomolecules* **2021**, *11*, 1345. [CrossRef]
46. Lindskog, M. Histamine Receptors in the Cross-Talk between Periphery and Brain. *Int. J. Neuropsychopharmacol.* **2017**, *20*, 400–402. [CrossRef]
47. Branco, A.C.C.C.; Yoshikawa, F.S.Y.; Pietrobon, A.J.; Sato, M.N. Role of Histamine in Modulating the Immune Response and Inflammation. *Mediat. Inflamm.* **2018**, *2018*, 9524075. [CrossRef]
48. Schofield, J.R.; Afrin, L.B. Recognition and Management of Medication Excipient Reactivity in Patients with Mast Cell Activation Syndrome. *Am. J. Med. Sci.* **2019**, *357*, 507–511. [CrossRef]
49. Haenisch, B.; Huber, M.; Wilhelm, T.; Steffens, M.; Molderings, G.J. Investigation into mechanisms mediating the inhibitory effect of 1,4-benzodiazepines on mast cells by gene expression profiling. *Life Sci.* **2013**, *92*, 345–351. [CrossRef]
50. Fukudo, S.; Kano, M.; Sato, Y.; Muratsubaki, T.; Kanazawa, M.; Tashiro, M.; Yanai, K. Histamine Neuroimaging in Stress-Related Disorders. In *The Functional Roles of Histamine Receptors*; Current Topics in Behavioral Neurosciences; 2022; Volume 59, pp. 113–129.
51. Panula, P. Histamine receptors, agonists, and antagonists in health and disease. *Handb. Clin. Neurol.* **2021**, *180*, 377–387. [CrossRef]
52. Xu, J.; Zhang, X.; Qian, Q.; Wang, Y.; Dong, H.; Li, N.; Qian, Y.; Jin, W. Histamine upregulates the expression of histamine receptors and increases the neuroprotective effect of astrocytes. *J. Neuroinflamm.* **2018**, *15*, 41. [CrossRef]
53. Lortholary, O.; Chandesris, M.O.; Livideanu, C.B.; Paul, C.; Guillet, G.; Jassem, E.; Niedoszytko, M.; Barete, S.; Verstovsek, S.; Grattan, C.; et al. Masitinib for treatment of severely symptomatic indolent systemic mastocytosis: A randomised, placebo-controlled, phase 3 study. *Lancet* **2017**, *389*, 612–620. [CrossRef] [PubMed]
54. Kohno, R.; Cannom, D.S.; Olshansky, B.; Xi, S.C.; Krishnappa, D.; Adkisson, W.O.; Norby, F.L.; Fedorowski, A.; Benditt, D.G. Mast Cell Activation Disorder and Postural Orthostatic Tachycardia Syndrome: A Clinical Association. *J. Am. Heart Assoc.* **2021**, *10*, e021002. [CrossRef] [PubMed]
55. Novak, P.; Giannetti, M.P.; Weller, E.; Hamilton, M.J.; Castells, M. Mast cell disorders are associated with decreased cerebral blood flow and small fiber neuropathy. *Ann. Allergy Asthma Immunol.* **2021**, *128*, 299–306.e1. [CrossRef] [PubMed]
56. Blitshteyn, S. Dysautonomia, Hypermobility Spectrum Disorders and Mast Cell Activation Syndrome as Migraine Comorbidities. *Curr. Neurol. Neurosci. Rep.* **2023**, *ahead of print*. [CrossRef]

Journal of *Personalized Medicine*

MDPI

Review

How Sex Hormones Affect Migraine: An Interdisciplinary Preclinical Research Panel Review

Frederick Godley III [1,*], John Meitzen [2], Hadas Nahman-Averbuch [3], Mary Angela O'Neal [4], David Yeomans [5], Nanette Santoro [6], Nina Riggins [7] and Lars Edvinsson [8]

1 Association of Migraine Disorders, P.O. Box 870, North Kingstown, RI 02852, USA
2 Department of Biological Sciences, NC State University, Raleigh, NC 27695, USA
3 Division of Clinical and Translational Research, Department of Anesthesiology, Washington University School of Medicine, St. Louis, MO 63110, USA
4 Brigham and Women's Hospital, 45 Francis St, Boston, MA 02115, USA
5 Department of Anesthesia, Pain and Perioperative Medicine, Stanford University School of Medicine, 300 Pasteur Drive, Stanford, CA 94305, USA
6 Department of Obstetrics and Gynecology, University of Colorado School of Medicine, Aurora, CO 80045, USA
7 Brain Performance Center and Research Institute, San Diego, CA 92122, USA
8 Division of Experimental Vascular Research, Department of Clinical Sciences, Lund University Hospital, 22185 Lund, Sweden
* Correspondence: rick@migrainedisorders.org

Abstract: Sex hormones and migraine are closely interlinked. Women report higher levels of migraine symptoms during periods of sex hormone fluctuation, particularly during puberty, pregnancy, and perimenopause. Ovarian steroids, such as estrogen and progesterone, exert complex effects on the peripheral and central nervous systems, including pain, a variety of special sensory and autonomic functions, and affective processing. A panel of basic scientists, when challenged to explain what was known about how sex hormones affect the nervous system, focused on two hormones: estrogen and oxytocin. Notably, other hormones, such as progesterone, testosterone, and vasopressin, are less well studied but are also highlighted in this review. When discussing what new therapeutic agent might be an alternative to hormone therapy and menopause replacement therapy for migraine treatment, the panel pointed to oxytocin delivered as a nasal spray. Overall, the conclusion was that progress in the preclinical study of hormones on the nervous system has been challenging and slow, that there remain substantial gaps in our understanding of the complex roles sex hormones play in migraine, and that opportunities remain for improved or novel therapeutic agents. Manipulation of sex hormones, perhaps through biochemical modifications where its positive effects are selected for and side effects are minimized, remains a theoretical goal, one that might have an impact on migraine disease and other symptoms of menopause. This review is a call to action for increased interest and funding for preclinical research on sex hormones, their metabolites, and their receptors. Interdisciplinary research, perhaps facilitated by a collaborative communication network or panel, is a possible strategy to achieve this goal.

Keywords: sex hormones; migraine; estrogen; oxytocin; progesterone; testosterone; prolactin; vasopressin

1. Introduction

Migraine is a neurological disorder affecting 12% of adults around the world at any one point in time [1]. Migraine symptoms can be different in women than men. Women can have more frequent and intense headaches with a higher risk of chronification [2]. Migraine is now recognized as the number one cause of disability globally for women aged 15–49 [3]. This gender difference in the behavior of migraine as a disease highlights the role of sex hormones in its pathophysiology. This review was the result of a round-table discussion among a panel of basic scientists from different disciplines on the topic of how sex hormones exert their effect on the nervous system, particularly migraine disease. The panelists were asked to discuss what the gaps in our knowledge were, what the barriers were, whether they could identify any new therapeutic agents that would provide an alternative treatment for migraine, and if there was an explanation for the clinical observation that the prevalence of some migraine-related symptoms, such as vestibular migraine and sinus pain and pressure, increase during perimenopause while headaches tend to recede.

2. Sex Hormone Fluctuation as a Trigger of Migraine

Migraine tends to follow a classic temporal pattern throughout a cisgender woman's life that corresponds with sex hormone fluctuations during reproductive milestones in the female lifespan. Puberty is a key period with significant changes in sex hormone levels. Interestingly, in children and adolescents, the prevalence of migraine headaches is nearly equivalent in boys and girls [4], but during puberty, the prevalence of migraine between men and women diverges and is 3–4 times higher in women compared to men [5,6]. This sex difference corresponds to the onset of menarche and falls after menopause.

Migraine symptoms can be linked to menstrual cycle changes (menstrual migraine) and 18–25% of women with migraine experience migraine or headaches during menstruation [7]. Menstrual migraine can be associated with a higher frequency of migraine-accompanying symptoms and more frequent and severe migraine attacks [8]. A comparison of women with and without migraine shows that those with migraine are characterized by faster late-luteal-phase estrogen decline compared to women without migraine. Thus, the timing and rate of estrogen withdrawal has been proposed to be a marker of vulnerability to migraine in women [9]. Contraceptive pills reduce the number of migraine attacks, migraine days, pain scores, disability scores, and migraine medication use while reducing the frequency of aura, and lowering, but not eliminating, the risks of cardiovascular complications or other side effects [10–12]. Another strategy is to use estrogen supplementation with a pill, vaginal gel or patch during the menstrual week.

Migraine is a heterogeneous disease associated with many possible combinations of genetic defects which share a common phenotype of intermittent pain or other hypersensitivities. This accounts for the unpredictable response of migraineurs to medications and the effect of hormones on the nociceptive system is no exception. For some, a drop in estrogen triggers a menstrual migraine attack without aura; for others, high levels of estrogen can trigger an attack with aura [13].

Migraine disease has a complex relationship with pregnancy. For 8% of women with migraine, their headaches worsen during the first trimester. This is especially true for migraine without aura, which is more hormonally driven [14–16]. The majority of women with migraine generally experience reduced migraine symptoms by the third trimester [17]. However, many women have the acute onset of headaches during pregnancy. Approximately 60% of these new headaches will be related to migraine but caution must be taken to evaluate pregnant women for secondary headaches [18]. A third of women will have postpartum headaches [19]. For those who continue to have migraine symptoms during their pregnancy and immediately postpartum, treatment options are limited to protect the fetus. There are specific recommendations for safe care of women with migraine headaches during pregnancy and breastfeeding [20].

Perimenopause, the period of two to eight years when menses first become irregular prior to the year after the end of menses, is a time when hormonal fluctuations are still occurring, and pre-existing migraine symptoms can remain unchanged, improve, or worsen [21–23]. In total, 8–13% of women report their first migraine during perimenopause [24,25]. However, many women see a decrease in headache prevalence during this period [26,27], most prominently in women who already suffer from migraine with aura [28]. For unexplained reasons, mid-facial pain and pressure and vestibular migraine can become prominent symptoms during perimenopause and menopause [29]. Hormone replacement therapy, or menopause replacement therapy (MRT), usually a continuous dosing of estrogen alone or estrogen plus progestin (ethinyl estradiol 5 μg combined with norethindrone acetate 1 mg, estradiol 1 mg combined with 0.5 mg norethindrone acetate, or transdermal estradiol combined with one-quarter or one-half of a 5 mg norethindrone daily) [30], remains an option, particularly for those women who have not had a hysterectomy because estrogen alone increases the risk of endometrial cancer. Transdermal estrogen patches or gels can be efficacious and less risky than systemic estrogen replacement in treating migraine [7,23,31]. A significant shortcoming of supplemental hormone therapies is that they do not provide migraine relief for all women and, for some, headaches become more severe. But a second major shortcoming of MRT is that, although the dosing of sex hormones is roughly half that of birth control pills, the risks of heart disease, stroke, blood clots, and breast cancer are not eliminated [13,30,32].

The bottom line is that current sex hormone supplements play a valuable role in mitigating the symptoms of migraine, but, because they are still associated with serious complications, especially migraine with aura, and exacerbate migraine symptoms in some, many medical professionals choose not to use hormone supplements in their migraine treatment plan. For example, plant-derived hormones (phytoestrogens) and the derivative bio-identical hormones are effective in reducing menstrual-related migraine headaches [33], but there is no rigorous scientific evidence that these supplements are safer or more natural compared to the current hormonal interventions. Phytoestrogen-containing foods, such as soy, are recommended over supplements, and all phytoestrogens should be avoided if there is a chance of pregnancy because these compounds might adversely affect the endocrine system. It is speculated that they might be safer in older women, such as those suffering from menopausal symptoms, particularly hot flashes [34,35], but currently there is not enough evidence to conclude that the benefits of phytoestrogens outweigh their potential health risks [36], and they do not appear to be ideal migraine preventive agents. Thus, since many women with migraine are unable to find an effective preventive therapy, there remains the challenge to understand how sex hormone supplements work, with the goal that select metabolites or synthetic derivatives might be both efficacious and safer than current hormonal therapies.

3. Which Sex Hormones Should Be the Target?

3.1. Estrogen

Estrogen plays a complicated role in migraine disease. Both drops and fluctuations in estrogen are associated with migraine symptoms, but its effect varies between individuals because of different receptors, metabolites, and interactions with other hormones. The dominant understanding of how crucial estrogen is in protecting individuals from migraine symptoms is what happens when estrogen levels decline: the estrogen withdrawal hypothesis. This hypothesis theorizes that drops in plasma estrogen trigger migraine attacks and neuroinflammation, eventually leading to chronic sensitization [37]. There are several possible mechanisms to explain his theory. One explanation is that estrogen suppresses pain by binding to estrogen receptor alpha (ER alpha) and estrogen receptor beta (ER beta), which are primarily associated with cell nuclei in the trigeminal ganglia. Activation of these nuclear receptors regulates inflammatory genes that ultimately suppresses cell excitability [38]. Also, this hypothesis may be explained by drops in estrogen leading to higher levels of calcitonin gene-related peptide (CGRP) [23].

CGRP is believed to be among the critical neuropeptides responsible for the throbbing pain associated with a migraine attack and the neuroinflammation that causes both pain and that perhaps cause neuroplastic neural changes responsible for chronic central sensitization [39]. Specifically, estrogen may also increase neurogenic vasodilation and gene regulation. For example, in mice, expression of neuropeptide Y and galanin, two neuropeptides which may inhibit or modulate CGRP mechanisms in trigeminal neurons, may play a part in the fluctuations of head pain during the estrus cycle [40].

While the estrogen withdrawal hypothesis focuses primarily on the trigeminal nerves, it is important to recognize the wider-ranging actions of estrogen in other parts of the body and brain [41]. A second mechanism to explain the estrogen withdrawal theory was demonstrated in an animal model where reduced levels of estrogen were shown to increase the frequency of cortical spreading depressions, the electrophysiological event believed to be responsible for triggering the trigeminal system and headaches, as well as auras [42].

There are various mechanisms that might explain how cortical spreading depressions are initiated. For example, estrogen is known to rapidly alter cellular excitability and gene expression in hypothalamic neurons [43,44]. And estrogen affects energy homeostasis via the proopiomelanocortin (POMC) neurons in the hypothalamic arcuate [45], and may play a role in migraine. Other brain regions, such as the mesolimbic cortical reward system, have also been implicated and show profound estrogen sensitivity [46–48]. The complexity stems from having three forms of estrogen (estrone, estradiol and estriol), thirteen estradiol metabolites, and two classes of receptors with different isomers which are functionally distinct and differentially distributed throughout the brain. Estrogen has other metabolic functions that might contribute to pain control indirectly, such as its indirect effect on serotonin [49].

3.2. Progesterone

Progesterone, the second major sex hormone, is produced in the ovaries, adrenal glands and placenta, and primarily helps maintain pregnancy. Progesterone with estradiol is found at the onset of menstrual migraines. Nonetheless, it is more likely that the withdrawal of estradiol, rather than progesterone, initiates migraine headaches. Instead, progesterone appears to protect neurons by suppressing neuroinflammation and reducing trigeminal nerve sensitivity. In one study, the receptive field size of facial trigeminal mechanoreceptors was not increased by treatment with progesterone, unlike the effects of estradiol [50].

It may be in the interplay with additional factors where progesterone plays an integral role in pain modulation. In a longitudinal study of fibromyalgia, it was high levels of progesterone and testosterone together that were associated with less pain [51]. Progesterone and testosterone are able to penetrate the blood-brain barrier and function as precursors for neurosteroids. There is an example of a progesterone derivative which enhances GABA function by modulating GABA receptors and, in turn, inhibits neuronal sensitivity [52,53]. Furthermore, both progesterone and allopregnanolone appear to dampen nociception in the trigeminovascular system and to reduce neurogenic inflammation in migraine through neuron-glia interactions [52]. In addition, in animals, progesterone and estradiol affect two CNS pathways that lead to increased neuroprotection [54]. But the role of progesterone in neuroinflammation is complicated by the finding that, during menstruation, prostaglandins rise and promote neuroinflammation through the release of substance P, neurokinins, and CGRP [55].

Currently, synthetic progesterone is used as a form of birth control and a migraine preventive agent in the form of a continuous low dose of progestin. Bio-identical progesterone can be delivered in three formulations: orally, topically, and as a suppository. Progesterone may improve insomnia as a mild sedative, and improve sleep apneas by stimulating respiration [56]. Finally, the progesterone metabolite, allopregnanolone, plays a role in the disproportionate level of mood disorders in susceptible women [57], and may begin to explain the high prevalence of anxiety in those with migraine.

3.3. Testosterone

A popular belief is that testosterone is the male hormone whereas estrogen is the female hormone. However, this is an oversimplification, as both estrogen and testosterone have important roles to play in individuals of either sex [58]. In both males and females, the balance between estrogen and testosterone production throughout life influences the function of both reproductive and nonreproductive organs [58].

Testosterone could be a potential therapeutic target, as it has an antinociceptive effect [59–63]. In animal studies, after gonadectomy or the blocking of testosterone receptors, animals appeared more sensitive to nociceptive stimuli [64–68]. The few human studies performed support an analgesic effect of testosterone, as higher testosterone levels are associated with lower experimental pain sensitivity [69]. Studies on the relationship of testosterone to migraine are few. Testosterone levels are lower in adults with migraine vs. without migraine, and are related to migraine severity. Interestingly, even when similar testosterone levels are found, men with migraine more frequently report symptoms of androgen deficiency compared to men with no migraine. However, one study found that no differences in testosterone levels were found in women with vs. without migraine, and that migraine pain intensity was not correlated with testosterone levels. In addition, transgender subjects who were given androgen-blocking medication and estrogen replacement developed increased levels of migraine with aura, similar to the effect of estrogen replacement therapy in cisgender women [13]. Since men with lower levels of androgen are prone to cluster headaches [70], the androgen deficiency model of migraine is based on the premise that testosterone offers neuroprotection. This theory is complicated by finding that, in contrast to estrogen which promotes neuroinflammation through CGRP and other neuropeptides, testosterone promotes neuroinflammation through microglial pathways. Therefore, while testosterone supplementation in females might protect against progression to chronic migraine, it will not have the same effect due to the gender-specific physiology of males [71].

Testosterone appears to be able to effectively reduce symptoms by suppressing spreading depressions, increasing serotonin, stabilizing cerebral blood flow, and reducing cell excitability and neuroinflammation [72]. These metabolic effects may explain the findings that testosterone treatment can improve clinical pain and experimental pain sensitivity in patients with chronic pain, including in patients with temporomandibular joint pain, fibromyalgia, and migraine [73–76], and that testosterone treatment delivered by a subcutaneous implant significantly reduces migraine intensity [75]. Thus, although testosterone is not thought to play a causal role in migraine, it likely modulates pain. Nonetheless, limited evidence and complex effects are reasons that testosterone is not included in migraine management guidelines.

3.4. Oxytocin

Oxytocin's (OT) therapeutic effects in migraine are complex and widespread in the nervous system, including at the level of the primary sensory neuron, spinal cord, and in a variety of brain regions associated with pain processing and modulation [77–79]. A recent theory is that menstrual migraines are related to a drop in both estrogen and OT during menstruation. Whether the lower concentrations of OT are secondary to the effect of less available estrogen in the CNS is not yet known.

The effect of OT on migraine has been shown via a case report in which intravenous OT provided analgesia and migraine relief [80]. In addition, double-blind, placebo-controlled clinical studies have shown evidence that intranasal OT sprays are efficacious for treating migraine pain in adult men and women [77,81] and experimental-evoked pain in men [82]. A benefit of oxytocin as a treatment for migraine is that it is routinely administered intranasally for inducing labor, postpartum care, and for enhancing lactation, and its safety profile is well documented. In addition, intranasal oxytocin in humans has no major side effects [83].

OT is a neuropeptide that exerts its pain-inhibitory effects both at the level of the primary afferent fiber and in the central nervous system. The first mechanism is via the descending neural pathway from the paraventricular nucleus (PVN) to the dorsal horn of the spinal cord [84,85]. Signals from the PVN release oxytocin in the spinal dorsal horn that activate GABAergic interneurons in the dorsal horn which secondarily recruit other inhibitory GABAergic interneurons and suppress pain signals carried by ascending A-delta and C-fibers [86–89]. The second mechanism is where OT released from the supraoptic nucleus (SON) in the hypothalamus, periaqueductal gray (PAG), rostral ventromedial medulla (RVM), and the spinal dorsal horn [90,91] modulates central endogenous pain pathways by raising nociceptive thresholds [92,93]. OT can suppress headache pain by binding to oxytocin receptors (OTRs) specifically in the trigeminal nucleus and trigeminal ganglia [94]. Imaging studies of migraine patients show overlap in the localization of OT/OTR, particularly those in the brainstem, thought to be migraine generators [95].

OTR mRNA and proteins are expressed in nociceptive C-fibers and Aδ-fibers in the adult rat trigeminal ganglia [94], and have a high level of co-expression with CGRP in trigeminal ganglia neurons [77]. OT dose-dependently blocks the release of calcitonin gene-related peptide (CGRP) from trigeminal afferent neurons innervating the dura in vitro [94]. CGRP is critical for the pathogenesis for chronic migraine, meaning that OTR activation on trigeminal nociceptive neurons could be a key mechanism of decreased headache intensity and frequency in migraine.

OT might have a general anti-inflammatory effect in orofacial nociceptive pathways by activating OTR, which can also suppress pro-inflammatory markers IL-1B and TNFa in the trigeminal ganglia (and in the spinal trigeminal nucleus caudalis) by inhibiting upregulation of these cytokines. A secondary effect is that inflammatory pain stimulates increased OTR gene expression [96]. But with less OT, trigeminal ganglia neurons become more sensitive, enhancing the likelihood of a migraine being triggered [97].

3.5. Vasopressin

Arginine vasopressin (AVP) is a neuropeptide hormone that has an antidiuretic effect in low concentrations, but at higher concentrations it causes vasoconstriction. Together, these effects raise blood pressure. AVP also has a role in pain, behavior, platelet aggregation, and blood coagulation functions. Specifically, AVP, in response to stress and pain, may be relevant to migraine pathophysiology [98,99]. Platelets have more AVP receptors in women who experience migraine [100]. It is possible that the AVP secretion has nothing directly to do with migraine, but, since the highest levels of AVP during a migraine attack may be associated with emesis [101] and vomiting, hypovolemia and nausea without vomiting trigger AVP release. Elevated levels of AVP may be responsible for the facial pallor, antidiuresis, and coagulation abnormalities occasionally observed in migraine [102]. In addition, some migraine precipitators (stress, ethanol, etc.) cause decreased AVP secretion and bioavailability, while some migraine-improving factors (tricyclic antidepressants, sleep, etc.) are associated with an increase in AVP [103]. Intranasal delivery of AVP has been described as an effective therapeutic agent for headache control [104].

Much of AVP is synthesized in the SON of the hypothalamus and, while AVP is largely stored in and secreted from the pituitary, AVP-containing hypothalamic fibers are widely distributed in the CNS [105]. These fibers reach different centers in the brainstem and, in particular, the trigemminal nuclei. The AVP receptors (VP1 and VP2) are found in the trigeminal ganglion [94]. Thus, the AVP system has many ways to modulate migraine pathophysiology. Since there are no direct fibers containing AVP in the trigeminovascular system, it is likely that the peptide may diffuse into this system. Overall, there exists sufficient evidence to maintain interest in the use of AVP to moderate the onset of headaches [106].

3.6. Prolactin

Prolactin (PRL) is a hormone that is responsible for lactation, breast development, and hundreds of other actions needed to maintain homeostasis. PRL is chemically related to growth hormones and placental lactogen hormones. In an animal model, high levels of prolactin increased meningeal trigeminal pain sensitivity by only affecting CGRP in female rodents [107]. In humans, serum prolactin levels are higher in those with migraine. Individuals with prolactin-secreting pituitary adenomas were found to have a higher incidence of headaches and migraine attacks [108]. With monoclonal antibodies targeting prolactin receptors, a recent report opens new possibilities to better understand the complex interaction between prolactin and CGRP, but blocking prolactin receptors in humans poses risks of interfering with the other functions of this hormone [109].

4. Limitations of Current Methods

Advances in understanding sex hormones in humans are hampered by the challenges of reliably creating an equivalent model of a migraine attack and measuring responses to interventions in animal models. Additionally, the translational value of preclinical studies can be uncertain due to a predominant use of males or not reporting sex as a biological variable [110,111], a reliance on ovariectomies, and modeling hormonal changes in animals that have an estrous cycle rather than a human-like menstrual cycle [112]. Furthermore, the effect of sex hormones on migraine and pain may vary depending on the pain model, model species, and experimental design in laboratory settings. The expert panel identified the lack of an established migraine animal model as one of the barriers to rapid progress in migraine research. For human research, the design of effective human studies has been challenging. Blood sampling of hormone levels is complicated by fluctuation throughout the day and month. The differential effect of sex hormone interventions might be impacted by the delivery method, timing of delivery, and dose, as well as sex, age and other conditions and medications of the patients.

As migraine is inherently a complex disorder involving different biological systems including the nervous, endocrine, endothelial, and immune systems, an interdisciplinary and collaborative approach among clinical and preclinical researchers is encouraged. Furthermore, given the limited number of basic scientists exploring this subject, it is critical that there is a cross-pollination of knowledge and ideas for research between often isolated fields of study. For example, chronic pain, which includes fibromyalgia, back pain, and TMJ overlaps with research performed in immunology, headache medicine, and other medical specialties [113,114]. It will take a dramatic increase and maintained effort in advocacy and support from patients and medical professionals to advance our knowledge of migraine and hormonal pathophysiology enough to lead to hormonal therapies of greater precision and safety.

5. Conclusions and Future Directions

While a large body of research has established hormonal changes and fluctuations as a driver of migraine symptoms in women and transgender people, the relation to hormonal life events is not definitively known for the full range of migraine symptoms. A new HEADS (headache, ear, auditory, dizziness, and sinus) Registry is now available to record and track many of these symptoms (reference: headsregistry.lumiio.com). Additionally, clarification of the mechanisms behind the emergence and recession of different migraine-related symptoms remains. Hormones may have both a causal role in migraine generation and also contribute to pain propagation.

This panel identified several potential hormones and mechanisms that show promise for improved migraine therapeutics, but the conclusion was that more resources need to be concentrated on this significantly debilitating neurovascular condition. In particular, the gender-specific nature of migraine disease calls for the need to better understand how hormones affect the nervous system. New areas of research are required to better understand the mechanisms by which sex hormones relate to changes in migraine symptoms during the periods of hormonal fluctuation in puberty, menstruation, pregnancy, and perimenopause. Translationally relevant animal models of migraine will play a key role in providing mechanistic insights, especially when coupled with clinical data. We highlight the theoretical opportunity to create novel hormone-based therapeutic molecules that might desensitize the hyperactive migraine nervous system without the potential side effects of contraceptives and hormone replacement therapy. Moreover, progress in understanding how hormones affect the nervous system will lead to innovations in treating not only migraine, but other menopausal symptoms.

Author Contributions: Conceptualization, F.G.III; writing—original draft preparation, F.G.III; writing—review and editing, all authors. All authors have read and agreed to the published version of the manuscript.

Funding: This research received no external funding.

Institutional Review Board Statement: Not applicable.

Informed Consent Statement: Not applicable.

Data Availability Statement: No new data were created or analyzed in this study. Data sharing is not applicable to this article.

Acknowledgments: Fred Schwaller, was a freelance writer who contributed to the manuscript; Leigh Serth was a valuable assistant in the manuscript preparation.

Conflicts of Interest: F.G. is a consultant for Allergen and Pfizer. A.O. is a consultant for Crico and Teladocs N.R. consulted for Gerson Lehrman Group, participated in compensated work with AcademicCME. was a Principal investigator (PI) on research with Electrocore, Theranica, Eli Lilly, was an uncompensated PI on research with products of Theraspecs, Dolor technologies, is an advisor for Theranica, is on the NeurologyLive Advisory board., is on the Board and received compensation for editing NeurologyLive issue, is a Board member of Miles for Migraine, and is a Project Advisor for Clinical Awareness Initiative with Clinical Neurological Society of America Inc. N.S. is a member of the Scientific Advisory Boards for Astellas and Menogenix, Inc, companies involved in clinical trials of non-hormone treatments for hot flashes, a consultant for Amazon Project Ember which is developing home measurements of hormones to be applied to women's health, and is a consultant for Ansh Laboratories, an immunoassay company that specializes in ovarian peptides. D. Y is an inventor of two oxytocin patents—one covering the use in headache, the other covering a magnesium formulation. These have been licensed by Tonix Pharmaceuticals which is pursuing a chronic migraine trial.

References

1. Burch, R.C.; Buse, D.C.; Lipton, R.B. Migraine: Epidemiology, Burden, and Comorbidity. *Neurol. Clin.* **2019**, *37*, 631–649. [CrossRef] [PubMed]
2. Tsai, C.K.; Tsai, C.L.; Lin, G.Y.; Yang, F.C.; Wang, S.J. Sex Differences in Chronic Migraine: Focusing on Clinical Features, Pathophysiology, and Treatments. *Curr. Pain Headache Rep.* **2022**, *26*, 347–355. [CrossRef] [PubMed]
3. Vos, T.; Abajobir, A.A.; Abate, K.H.; Abbafati, C.; Abbas, K.M.; Abd-Allah, F.; Abdulkader, R.S.; Abdulle, A.M.; Abebo, T.A.; Abera, S.F.; et al. Global, regional, and national incidence, prevalence, and years lived with disability for 328 diseases and injuries for 195 countries, 1990–2016: A systematic analysis for the Global Burden of Disease Study 2016. *Lancet* **2017**, *390*, 1211–1259. [CrossRef] [PubMed]
4. Szperka, C. Headache in Children and Adolescents. *Continuum* **2021**, *27*, 703–731. [CrossRef] [PubMed]
5. Tonini, M.C. Gender differences in migraine. *Neurol. Sci.* **2018**, *39*, 77–78. [CrossRef] [PubMed]
6. Vetvik, K.G.; MacGregor, E.A. Sex differences in the epidemiology, clinical features, and pathophysiology of migraine. *Lancet Neurol.* **2017**, *16*, 76–87. [CrossRef]
7. MacGregor, E.A.; Frith, A.; Ellis, J.; Aspinall, L.; Hackshaw, A. Incidence of migraine relative to menstrual cycle phases of rising and falling estrogen. *Neurology* **2006**, *67*, 2154–2158. [CrossRef] [PubMed]
8. Chalmer, M.A.; Kogelman, L.J.A.; Ullum, H.; Sørensen, E.; Didriksen, M.; Mikkelsen, S.; Dinh, K.M.; Brodersen, T.; Nielsen, K.R.; Bruun, M.T.; et al. Population-Based Characterization of Menstrual Migraine and Proposed Diagnostic Criteria. *JAMA Netw. Open* **2023**, *6*, e2313235. [CrossRef]
9. Pavlović, J.M.; Allshouse, A.A.; Santoro, N.F.; Crawford, S.L.; Thurston, R.C.; Neal-Perry, G.S.; Lipton, R.B.; Derby, C.A. Sex hormones in women with and without migraine: Evidence of migraine-specific hormone profiles. *Neurology* **2016**, *87*, 49–56. [CrossRef]
10. Calhoun, A.H.; Batur, P. Combined hormonal contraceptives and migraine: An update on the evidence. *Clevel. Clin. J. Med.* **2017**, *84*, 631–638. [CrossRef]
11. Warhurst, S.; Rofe, C.J.; Brew, B.J.; Bateson, D.; McGeechan, K.; Merki-Feld, G.S.; Garrick, R.; Tomlinson, S.E. Effectiveness of the progestin-only pill for migraine treatment in women: A systematic review and meta-analysis. *Cephalalgia* **2018**, *38*, 754–764. [CrossRef] [PubMed]
12. Lyall, M.; de Oliveira, B.R.; Mody, S.K. Considerations for Contraceptive Use Among Patients with Migraines. *Curr. Obstet. Gynecol. Rep.* **2023**, *12*, 57–63. [CrossRef]
13. MacGregor, E.A. Migraine, menopause and hormone replacement therapy. *Post Reprod. Health* **2018**, *24*, 11–18. [CrossRef]
14. Aubé, M. Migraine in pregnancy. *Neurology* **1999**, *53* (Suppl. S1), S26–S28.

15. Ertresvåg, J.M.; Zwart, J.A.; Helde, G.; Johnsen, H.J.; Bovim, G. Headache and transient focal neurological symptoms during pregnancy, a prospective cohort. *Acta Neurol. Scand.* **2005**, *111*, 233–237. [CrossRef]
16. Kvisvik, E.V.; Stovner, L.J.; Helde, G.; Bovim, G.; Linde, M. Headache and migraine during pregnancy and puerperium: The MIGRA-study. *J. Headache Pain* **2011**, *12*, 443–451. [CrossRef]
17. Goadsby, P.J.; Goldberg, J.; Silberstein, S.D. Migraine in pregnancy. *BMJ* **2008**, *336*, 1502–1504. [CrossRef] [PubMed]
18. Robbins, M.S.; Farmakidis, C.; Dayal, A.K.; Lipton, R.B. Acute headache diagnosis in pregnant women: A hospital-based study. *Neurology* **2015**, *85*, 1024–1030. [CrossRef]
19. Sances, G.; Granella, F.; Nappi, R.E.; Fignon, A.; Ghiotto, N.; Polatti, F.; Nappi, G. Course of migraine during pregnancy and postpartum: A prospective study. *Cephalalgia* **2003**, *23*, 197–205. [CrossRef]
20. Wells, R.E.; Turner, D.P.; Lee, M.; Bishop, L.; Strauss, L. Managing Migraine during Pregnancy and Lactation. *Curr. Neurol. Neurosci. Rep.* **2016**, *16*, 40. [CrossRef]
21. Mueller, L. Predictability of exogenous hormone effect on subgroups of migraineurs. *Headache* **2000**, *40*, 189–193. [CrossRef] [PubMed]
22. Hodson, J.; Thompson, J.; Al-Azzawi, F. Headache at menopause and in hormone replacement therapy users. *Climacteric J. Int. Menopause Soc.* **2000**, *3*, 119–124. [CrossRef] [PubMed]
23. Ibrahimi, K.; Couturier, E.G.M.; MaassenVanDenBrink, A. Migraine and perimenopause. *Maturitas* **2014**, *78*, 277–280. [CrossRef] [PubMed]
24. Granella, F.; Sances, G.; Zanferrari, C.; Costa, A.; Martignoni, E.; Manzoni, G.C. Migraine without aura and reproductive life events: A clinical epidemiological study in 1300 women. *Headache* **1993**, *33*, 385–389. [CrossRef] [PubMed]
25. Cupini, L.M.; Matteis, M.; Troisi, E.; Calabresi, P.; Bernardi, G.; Silvestrini, M. Sex-hormone-related events in migrainous females. A clinical comparative study between migraine with aura and migraine without aura. *Cephalalgia* **1995**, *15*, 140–144. [CrossRef] [PubMed]
26. Wang, S.J.; Fuh, J.L.; Lu, S.R.; Juang, K.D.; Wang, P.H. Migraine prevalence during menopausal transition. *Headache* **2003**, *43*, 470–478. [CrossRef] [PubMed]
27. Freeman, E.W.; Sammel, M.D.; Lin, H.; Gracia, C.R.; Kapoor, S. Symptoms in the menopausal transition: Hormone and behavioral correlates. *Obstet. Gynecol.* **2008**, *111*, 127–136. [CrossRef]
28. Mattsson, P. Hormonal factors in migraine: A population-based study of women aged 40 to 74 years. *Headache* **2003**, *43*, 27–35. [CrossRef]
29. Park, J.H.; Viirre, E. Vestibular migraine may be an important cause of dizziness/vertigo in perimenopausal period. *Med. Hypotheses* **2010**, *75*, 409–414. [CrossRef]
30. Pavlović, J.M. Evaluation and management of migraine in midlife women. *Menopause* **2018**, *25*, 927–929. [CrossRef]
31. Kaiser, H.J.; Meienberg, O. Deterioration or onset of migraine under oestrogen replacement therapy in the menopause. *J. Neurol.* **1993**, *240*, 195–196. [CrossRef]
32. Sacco, S.; Ricci, S.; Degan, D.; Carolei, A. Migraine in women: The role of hormones and their impact on vascular diseases. *J. Headache Pain* **2012**, *13*, 177–189. [CrossRef]
33. Burke, B.E.; Olson, R.D.; Cusack, B.J. Randomized, controlled trial of phytoestrogen in the prophylactic treatment of menstrual migraine. *Biomed. Pharmacother.* **2002**, *56*, 283–288. [CrossRef] [PubMed]
34. Patisaul, H.B.; Jefferson, W. The pros and cons of phytoestrogens. *Front. Neuroendocrinol.* **2010**, *31*, 400–419. [CrossRef]
35. Chen, M.N.; Lin, C.C.; Liu, C.F. Efficacy of phytoestrogens for menopausal symptoms: A meta-analysis and systematic review. *Climacteric* **2015**, *18*, 260–269. [CrossRef]
36. Rietjens, I.M.C.M.; Louisse, J.; Beekmann, K. The potential health effects of dietary phytoestrogens. *Br. J. Pharmacol.* **2017**, *174*, 1263–1280. [CrossRef]
37. Sarajari, S.; Oblinger, M.M. Estrogen Effects on Pain Sensitivity and Neuropeptide Expression in Rat Sensory Neurons. *Exp. Neurol.* **2010**, *224*, 163–169. [CrossRef] [PubMed]
38. Welch, K.M.A.; Brandes, J.L.; Berman, N.E.J. Mismatch in how oestrogen modulates molecular and neuronal function may explain menstrual migraine. *Neurol. Sci.* **2006**, *27* (Suppl. S2), S190–S192. [CrossRef] [PubMed]
39. Wattiez, A.S.; Sowers, L.P.; Russo, A.F. Calcitonin gene-related peptide (CGRP): Role in migraine pathophysiology and therapeutic targeting. *Expert Opin. Ther. Targets* **2020**, *24*, 91–100. [CrossRef] [PubMed]
40. Puri, V.; Cui, L.; Liverman, C.S.; Roby, K.F.; Klein, R.M.; Welch, K.M.A.; Berman, N.E.J. Ovarian steroids regulate neuropeptides in the trigeminal ganglion. *Neuropeptides* **2005**, *39*, 409–417. [CrossRef]
41. Rettberg, J.R.; Yao, J.; Brinton, R.D. Estrogen: A master regulator of bioenergetic systems in the brain and body. *Front. Neuroendocrinol.* **2014**, *35*, 8–30. [CrossRef] [PubMed]
42. Kudo, C.; Harriott, A.M.; Moskowitz, M.A.; Waeber, C.; Ayata, C. Estrogen modulation of cortical spreading depression. *J. Headache Pain* **2023**, *24*, 62. [CrossRef] [PubMed]
43. Chen, Q.; Zhang, W.; Sadana, N.; Chen, X. Estrogen receptors in pain modulation: Cellular signaling. *Biol. Sex Differ.* **2021**, *12*, 22. [CrossRef] [PubMed]
44. Kelly, M.J.; Rønnekleiv, O.K. Minireview: Neural Signaling of Estradiol in the Hypothalamus. *Mol. Endocrinol.* **2015**, *29*, 645–657. [CrossRef]

45. Stincic, T.L.; Grachev, P.; Bosch, M.A.; Rønnekleiv, O.K.; Kelly, M.J. Estradiol Drives the Anorexigenic Activity of Proopiomelanocortin Neurons in Female Mice. *eNeuro* **2018**, *5*, ENEURO.0103-18.2018. [CrossRef]
46. Cahill, C.M.; Cook, C.; Pickens, S. Migraine and Reward System—Or Is It Aversive? *Curr. Pain Headache Rep.* **2014**, *18*, 410. [CrossRef]
47. Krentzel, A.A.; Proaño, S.B.; Dorris, D.M.; Setzer, B.; Meitzen, J. The estrous cycle and 17β-estradiol modulate the electrophysiological properties of rat nucleus accumbens core medium spiny neurons. *J. Neuroendocrinol.* **2022**, *34*, e13122. [CrossRef]
48. Yoest, K.E.; Quigley, J.A.; Becker, J.B. Rapid effects of ovarian hormones in dorsal striatum and nucleus accumbens. *Horm. Behav.* **2018**, *104*, 119–129. [CrossRef]
49. Aggarwal, M.; Puri, V.; Puri, S. Effects of estrogen on the serotonergic system and calcitonin gene-related peptide in trigeminal ganglia of rats. *Ann. Neurosci.* **2012**, *19*, 151–157. [CrossRef] [PubMed]
50. Bereiter, D.A.; Stanford, L.R.; Barker, D.J. Hormone-induced enlargement of receptive fields in trigeminal mechanoreceptive neurons. II. possible mechanisms. *Brain Res.* **1980**, *184*, 411–423. [CrossRef] [PubMed]
51. Schertzinger, M.; Wesson-Sides, K.; Parkitny, L.; Younger, J. Daily Fluctuations of Progesterone and Testosterone Are Associated with Fibromyalgia Pain Severity. *J. Pain* **2018**, *19*, 410–417. [CrossRef]
52. Chuang, S.H.; Reddy, D.S. 3β-Methyl-Neurosteroid Analogs Are Preferential Positive Allosteric Modulators and Direct Activators of Extrasynaptic δ-Subunit γ-Aminobutyric Acid Type A Receptors in the Hippocampus Dentate Gyrus Subfield. *J. Pharmacol. Exp. Ther.* **2018**, *365*, 583–601. [CrossRef]
53. Reddy, D.S. Neurosteroids: Endogenous role in the human brain and therapeutic potentials. *Prog. Brain Res.* **2010**, *186*, 113–137. [CrossRef] [PubMed]
54. Singh, M. Ovarian Hormones Elicit Phosphorylation of Akt and Extracellular-Signal Regulated Kinase in Explants of the Cerebral Cortex. *ENDO* **2001**, *14*, 407–416. [CrossRef] [PubMed]
55. Jang, Y.; Kim, M.; Hwang, S.W. Molecular mechanisms underlying the actions of arachidonic acid-derived prostaglandins on peripheral nociception. *J. Neuroinflamm.* **2020**, *17*, 30. [CrossRef] [PubMed]
56. Andersen, M.L.; Bittencourt, L.R.A.; Antunes, B.I.; Tufik, S. Effects of Progesterone on Sleep: A Possible Pharmacological Treatment for Sleep-Breathing Disorders? *CMC* **2006**, *13*, 3575–3582. [CrossRef]
57. Standeven, L.R.; McEvoy, K.O.; Osborne, L.M. Progesterone, reproduction, and psychiatric illness. *Best Pract. Res. Clin. Obstet. Gynaecol.* **2020**, *69*, 108–126. [CrossRef]
58. Simpson, E.R. Sources of estrogen and their importance. *J. Steroid Biochem. Mol. Biol.* **2003**, *86*, 225–230. [CrossRef]
59. Bartley, E.J.; Palit, S.; Kuhn, B.L.; Kerr, K.L.; Terry, E.L.; DelVentura, J.L.; Rhudy, J.L. Nociceptive processing in women with premenstrual dysphoric disorder (PMDD): The role of menstrual phase and sex hormones. *Clin. J. Pain* **2015**, *31*, 304–314. [CrossRef]
60. Bartley, E.J.; Palit, S.; Kuhn, B.L.; Kerr, K.L.; Terry, E.L.; DelVentura, J.L.; Rhudy, J.L. Natural variation in testosterone is associated with hypoalgesia in healthy women. *Clin. J. Pain* **2015**, *31*, 730–739. [CrossRef]
61. Choi, J.C.; Park, Y.H.; Park, S.K.; Lee, J.S.; Kim, J.; Choi, J.I.; Yoon, K.B.; Lee, S.; Lim, D.E.; Choi, J.Y.; et al. Testosterone effects on pain and brain activation patterns. *Acta Anaesthesiol. Scand.* **2017**, *61*, 668–675. [CrossRef]
62. Choi, J.C.; Chung, M.I.; Lee, Y.D. Modulation of pain sensation by stress-related testosterone and cortisol. *Anaesthesia* **2012**, *67*, 1146–1151. [CrossRef] [PubMed]
63. Teepker, M.; Peters, M.; Vedder, H.; Schepelmann, K.; Lautenbacher, S. Menstrual variation in experimental pain: Correlation with gonadal hormones. *Neuropsychobiology* **2010**, *61*, 131–140. [CrossRef]
64. Aloisi, A.M.; Ceccarelli, I.; Fiorenzani, P. Gonadectomy affects hormonal and behavioral responses to repetitive nociceptive stimulation in male rats. *Ann. N. Y. Acad. Sci.* **2003**, *1007*, 232–237. [CrossRef] [PubMed]
65. Ceccarelli, I.; Scaramuzzino, A.; Massafra, C.; Aloisi, A.M. The behavioral and neuronal effects induced by repetitive nociceptive stimulation are affected by gonadal hormones in male rats. *Pain* **2003**, *104*, 35–47. [CrossRef] [PubMed]
66. Gaumond, I.; Arsenault, P.; Marchand, S. Specificity of female and male sex hormones on excitatory and inhibitory phases of formalin-induced nociceptive responses. *Brain Res.* **2005**, *1052*, 105–111. [CrossRef] [PubMed]
67. Aloisi, A.M.; Ceccarelli, I.; Fiorenzani, P.; De Padova, A.M.; Massafra, C. Testosterone affects formalin-induced responses differently in male and female rats. *Neurosci. Lett.* **2004**, *361*, 262–264. [CrossRef] [PubMed]
68. Stoffel, E.C.; Ulibarri, C.M.; Craft, R.M. Gonadal steroid hormone modulation of nociception, morphine antinociception and reproductive indices in male and female rats. *Pain* **2003**, *103*, 285–302. [CrossRef]
69. Basaria, S.; Travison, T.G.; Alford, D.; Knapp, P.E.; Teeter, K.; Cahalan, C.; Eder, R.; Lakshman, K.; Bachman, E.; Mensing, G.; et al. Effects of testosterone replacement in men with opioid-induced androgen deficiency: A randomized controlled trial. *Pain* **2015**, *156*, 280–288. [CrossRef]
70. Verhagen, I.E.; Brandt, R.B.; Kruitbosch, C.M.A.; MaassenVanDenBrink, A.; Fronczek, R.; Terwindt, G.M. Clinical symptoms of androgen deficiency in men with migraine or cluster headache: A cross-sectional cohort study. *J. Headache Pain* **2021**, *22*, 125. [CrossRef]
71. Martinez, C.I.; Liktor-Busa, E.; Largent-Milnes, T.M. Molecular mechanisms of hormones implicated in migraine and the translational implication for transgender patients. *Front. Pain Res.* **2023**, *4*, 1117842. [CrossRef]
72. Ahmad, A.H.; Ismail, Z. c-fos and its Consequences in Pain. *Malays. J. Med. Sci.* **2002**, *9*, 3–8. [PubMed]

73. White, H.D.; Brown, L.A.J.; Gyurik, R.J.; Manganiello, P.D.; Robinson, T.D.; Hallock, L.S.; Lewis, L.D.; Yeo, Y.-T.J. Treatment of pain in fibromyalgia patients with testosterone gel: Pharmacokinetics and clinical response. *Int. Immunopharmacol.* **2015**, *27*, 249–256. [CrossRef] [PubMed]
74. Fischer, L.; Clemente, J.T.; Tambeli, C.H. The protective role of testosterone in the development of temporomandibular joint pain. *J. Pain* **2007**, *8*, 437–442. [CrossRef] [PubMed]
75. Glaser, R.; Dimitrakakis, C.; Trimble, N.; Martin, V. Testosterone pellet implants and migraine headaches: A pilot study. *Maturitas* **2012**, *71*, 385–388. [CrossRef] [PubMed]
76. English, K.M.; Steeds, R.P.; Jones, T.H.; Diver, M.J.; Channer, K.S. Low-dose transdermal testosterone therapy improves angina threshold in men with chronic stable angina: A randomized, double-blind, placebo-controlled study. *Circulation* **2000**, *102*, 1906–1911. [CrossRef] [PubMed]
77. Tzabazis, A.; Kori, S.; Mechanic, J.; Miller, J.; Pascual, C.; Manering, N.; Carson, D.; Klukinov, M.; Spierings, E.; Jacobs, D.; et al. Oxytocin and Migraine Headache. *Headache* **2017**, *57* (Suppl. S2), 64–75. [CrossRef] [PubMed]
78. Rash, J.A.; Aguirre-Camacho, A.; Campbell, T.S. Oxytocin and pain: A systematic review and synthesis of findings. *Clin. J. Pain* **2014**, *30*, 453–462. [CrossRef] [PubMed]
79. Shamay-Tsoory, S.G.; Abu-Akel, A. The Social Salience Hypothesis of Oxytocin. *Biol. Psychiatry* **2016**, *79*, 194–202. [CrossRef]
80. Phillips, W.J.; Ostrovsky, O.; Galli, R.L.; Dickey, S. Relief of acute migraine headache with intravenous oxytocin: Report of two cases. *J. Pain Palliat. Care Pharmacother.* **2006**, *20*, 25–28. [CrossRef]
81. Wang, Y.L.; Yuan, Y.; Yang, J.; Wang, C.H.; Pan, Y.J.; Lu, L.; Wu, Y.Q.; Wang, D.X.; Lv, L.X.; Li, R.R.; et al. The interaction between the oxytocin and pain modulation in headache patients. *Neuropeptides* **2013**, *47*, 93–97. [CrossRef]
82. Paloyelis, Y.; Krahé, C.; Maltezos, S.; Williams, S.C.; Howard, M.A.; Fotopoulou, A. The Analgesic Effect of Oxytocin in Humans: A Double-Blind, Placebo-Controlled Cross-Over Study Using Laser-Evoked Potentials. *J. Neuroendocrinol.* **2016**, *28*. [CrossRef] [PubMed]
83. MacDonald, E.; Dadds, M.R.; Brennan, J.L.; Williams, K.; Levy, F.; Cauchi, A.J. A review of safety, side-effects and subjective reactions to intranasal oxytocin in human research. *Psychoneuroendocrinology* **2011**, *36*, 1114–1126. [CrossRef] [PubMed]
84. Iwasaki, M.; Lefevre, A.; Althammer, F.; Creusot, E.C.; Łąpieś, O.; Petitjean, H.; Hilfiger, L.; Kerspern, D.; Melchior, M.; Küppers, S.; et al. An analgesic pathway from parvocellular oxytocin neurons to the periaqueductal gray in rats. *Nat. Commun.* **2023**, *14*, 1066. [CrossRef] [PubMed]
85. Swanson, L.W.; McKellar, S. The distribution of oxytocin- and neurophysin-stained fibers in the spinal cord of the rat and monkey. *J. Comp. Neurol.* **1979**, *188*, 87–106. [CrossRef] [PubMed]
86. Breton, J.D.; Veinante, P.; Uhl-Bronner, S.; Vergnano, A.M.; Freund-Mercier, M.J.; Schlichter, R.; Poisbeau, P. Oxytocin-induced antinociception in the spinal cord is mediated by a subpopulation of glutamatergic neurons in lamina I-II which amplify GABAergic inhibition. *Mol. Pain* **2008**, *4*, 19. [CrossRef] [PubMed]
87. Rojas-Piloni, G.; López-Hidalgo, M.; Martínez-Lorenzana, G.; Rodríguez-Jiménez, J.; Condés-Lara, M. GABA-mediated oxytocinergic inhibition in dorsal horn neurons by hypothalamic paraventricular nucleus stimulation. *Brain Res.* **2007**, *1137*, 69–77. [CrossRef] [PubMed]
88. Condés-Lara, M.; González, N.M.; Martínez-Lorenzana, G.; Delgado, O.L.; Freund-Mercier, M.J. Actions of oxytocin and interactions with glutamate on spontaneous and evoked dorsal spinal cord neuronal activities. *Brain Res.* **2003**, *976*, 75–81. [CrossRef]
89. Miranda-Cardenas, Y.; Rojas-Piloni, G.; Martínez-Lorenzana, G.; Rodríguez-Jiménez, J.; López-Hidalgo, M.; Freund-Mercier, M.J.; Condés-Lara, M. Oxytocin and electrical stimulation of the paraventricular hypothalamic nucleus produce antinociceptive effects that are reversed by an oxytocin antagonist. *Pain* **2006**, *122*, 182–189. [CrossRef]
90. Yang, J.; Liang, J.Y.; Li, P.; Pan, Y.J.; Qiu, P.Y.; Zhang, J.; Hao, F.; Wang, D.X. Oxytocin in the periaqueductal gray participates in pain modulation in the rat by influencing endogenous opiate peptides. *Peptides* **2011**, *32*, 1255–1261. [CrossRef]
91. Yang, J.; Li, P.; Liang, J.Y.; Pan, Y.J.; Yan, X.Q.; Yan, F.L.; Hao, F.; Zhang, X.Y.; Zhang, J.; Qiu, P.Y.; et al. Oxytocin in the periaqueductal grey regulates nociception in the rat. *Regul. Pept.* **2011**, *169*, 39–42. [CrossRef]
92. Yang, J.; Yang, Y.; Chen, J.M.; Liu, W.Y.; Wang, C.H.; Lin, B.C. Central oxytocin enhances antinociception in the rat. *Peptides* **2007**, *28*, 1113–1119. [CrossRef] [PubMed]
93. Yang, J.; Yang, Y.; Chen, J.M.; Liu, W.Y.; Lin, B.C. Investigating the role of the hypothalamic supraoptic nucleus in nociception in the rat. *Life Sci.* **2008**, *82*, 166–173. [CrossRef] [PubMed]
94. Warfvinge, K.; Krause, D.N.; Maddahi, A.; Grell, A.-S.; Edvinsson, J.C.A.; Haanes, K.A.; Edvinsson, L. Oxytocin as a regulatory neuropeptide in the trigeminovascular system: Localization, expression and function of oxytocin and oxytocin receptors. *Cephalalgia* **2020**, *40*, 1283–1295. [CrossRef] [PubMed]
95. Warfvinge, K.; Krause, D.; Edvinsson, L. The distribution of oxytocin and the oxytocin receptor in rat brain: Relation to regions active in migraine. *J. Headache Pain* **2020**, *21*, 10. [CrossRef] [PubMed]
96. Huang, C.L.; Liu, F.; Zhang, Y.Y.; Lin, J.; Fu, M.; Li, Y.L.; Zhou, C.; Li, C.J.; Shen, J.F. Activation of oxytocin receptor in the trigeminal ganglion attenuates orofacial ectopic pain attributed to inferior alveolar nerve injury. *J. Neurophysiol.* **2021**, *125*, 223–231. [CrossRef] [PubMed]
97. Bharadwaj, V.N.; Porreca, F.; Cowan, R.P.; Kori, S.; Silberstein, S.D.; Yeomans, D.C. A new hypothesis linking oxytocin to menstrual migraine. *Headache J. Head Face Pain* **2021**, *61*, 1051–1059. [CrossRef] [PubMed]

98. Dalkara, T.; Nozari, A.; Moskowitz, M.A. Migraine aura pathophysiology: The role of blood vessels and microembolisation. *Lancet Neurol.* **2010**, *9*, 309–317. [CrossRef] [PubMed]
99. Gupta, V. Does Vasopressin Serve a Vasomotor Adaptive Function in Migraine? *Cephalalgia* **1993**, *13*, 221. [CrossRef]
100. Buschmann, J.; Leppla-Wollsiffer, G.; Nemeth, N.; Nelson, K.; Kirsten, R. Migraine patients show increased platelet vasopressin receptors. *Headache* **1996**, *36*, 586–588. [CrossRef]
101. Hampton, K.K.; Esack, A.; Peatfield, R.C.; Grant, P.J. Elevation of plasma vasopressin in spontaneous migraine. *Cephalalgia* **1991**, *11*, 249–250. [CrossRef]
102. Peatfield, R.C.; Hampton, K.K.; Grant, P.J. Plasma vasopressin levels in induced migraine attacks. *Cephalalgia* **1988**, *8*, 55–57. [CrossRef]
103. Bahadoram, M.; Mahmoudian-Sani, M.R.; Keikhaei, B.; Alikhani, K.; Bahadoram, S. The antimigraine action of arginine-vasopressin: A theoretical basis. *Future Neurol.* **2020**, *15*, FNL51. [CrossRef]
104. Yang, J.; Lu, L.; Wang, H.C.; Zhan, H.Q.; Hai, G.F.; Pan, Y.J.; Lv, Q.Q.; Wang, D.X.; Wu, Y.Q.; Li, R.R.; et al. Effect of intranasal arginine vasopressin on human headache. *Peptides* **2012**, *38*, 100–104. [CrossRef] [PubMed]
105. Warfvinge, K.; Krause, D.N.; Maddahi, A.; Edvinsson, J.C.A.; Edvinsson, L.; Haanes, K.A. Estrogen receptors α, β and GPER in the CNS and trigeminal system—Molecular and functional aspects. *J. Headache Pain* **2020**, *21*, 131. [CrossRef]
106. Maddahi, A.; Edvinsson, L.; Warfvinge, K. Expression of vasopressin and its receptors in migraine-related regions in CNS and the trigeminal system: Influence of sex. *J. Headache Pain* **2022**, *23*, 152. [CrossRef]
107. Avona, A.; Mason, B.N.; Burgos-Vega, C.; Hovhannisyan, A.H.; Belugin, S.N.; Mecklenburg, J.; Goffin, V.; Wajahat, N.; Price, T.J.; Akopian, A.N.; et al. Meningeal CGRP-Prolactin Interaction Evokes Female-Specific Migraine Behavior. *Ann. Neurol.* **2021**, *89*, 1129–1144. [CrossRef] [PubMed]
108. Al-Karagholi, M.A.M.; Kalatharan, V.; Ghanizada, H.; Gram, C.; Dussor, G.; Ashina, M. Prolactin in headache and migraine: A systematic review of clinical studies. *Cephalalgia* **2023**, *43*, 3331024221136286. [CrossRef]
109. Maciuba, S.; Bowden, G.D.; Stratton, H.J.; Wisniewski, K.; Schteingart, C.D.; Almagro, J.C.; Valadon, P.; Lowitz, J.; Glaser, S.M.; Lee, G.; et al. Discovery and characterization of prolactin neutralizing monoclonal antibodies for the treatment of female-prevalent pain disorders. *mAbs* **2023**, *15*, 2254676. [CrossRef] [PubMed]
110. Mamlouk, G.M.; Dorris, D.M.; Barrett, L.R.; Meitzen, J. Sex bias and omission in neuroscience research is influenced by research model and journal, but not reported NIH funding. *Front. Neuroendocrinol.* **2020**, *57*, 100835. [CrossRef]
111. Will, T.R.; Proaño, S.B.; Thomas, A.M.; Kunz, L.M.; Thompson, K.C.; Ginnari, L.A.; Jones, C.H.; Lucas, S.-C.; Reavis, E.M.; Dorris, D.M.; et al. Problems and Progress regarding Sex Bias and Omission in Neuroscience Research. *eNeuro* **2017**, *4*, ENEURO.0278-17.2017. [CrossRef] [PubMed]
112. Labastida-Ramírez, A.; Rubio-Beltrán, E.; Villalón, C.M.; MaassenVanDenBrink, A. Gender aspects of CGRP in migraine. *Cephalalgia* **2019**, *39*, 435–444. [CrossRef] [PubMed]
113. Lenert, M.E.; Avona, A.; Garner, K.M.; Barron, L.R.; Burton, M.D. Sensory Neurons, Neuroimmunity, and Pain Modulation by Sex Hormones. *Endocrinology* **2021**, *162*, bqab109. [CrossRef]
114. Harvey, M.P.; Dubois, M.C.; Chalaye, P.; Sansoucy, Y.; Marchand, S. Sex-Related Effects of Adrenergic Drugs on Conditioned Pain Modulation: A Randomized Controlled Cross-Over Double-Blind Trial. *Pain Res. Manag.* **2022**, *2022*, 2757101. [CrossRef] [PubMed]

Journal of
Personalized Medicine

MDPI

Review

Long COVID Is Not a Functional Neurologic Disorder

Todd E. Davenport [1,2,*], Svetlana Blitshteyn [3,4], Nicola Clague-Baker [5], David Davies-Payne [6], Glenn J. Treisman [7] and Sarah F. Tyson [8]

1 Department of Physical Therapy, University of the Pacific, Stockton, CA 95211, USA
2 Workwell Foundation, Santa Rosa, CA 95403, USA
3 Department of Neurology, Jacobs School of Medicine and Biomedical Sciences, University of Buffalo, Buffalo, NY 14203, USA
4 Dysautonomia Clinic, Williamsville, NY 14221, USA
5 School of Allied Health Professions and Nursing, Institute of Population Health, University of Liverpool, Liverpool L69 7ZX, UK
6 Department of Radiology, Starship Children's Hospital, Auckland 1023, New Zealand
7 Department of Psychiatry, Johns Hopkins School of Medicine, Baltimore, MD 21287, USA; glenn@jhmi.edu
8 School of Health Sciences, University of Manchester, Manchester M14 4PX, UK; sarah.tyson@manchester.ac.uk
* Correspondence: tdavenport@pacific.edu

Abstract: Long COVID is a common sequela of SARS-CoV-2 infection. Data from numerous scientific studies indicate that long COVID involves a complex interaction between pathophysiological processes. Long COVID may involve the development of new diagnosable health conditions and exacerbation of pre-existing health conditions. However, despite this rapidly accumulating body of evidence regarding the pathobiology of long COVID, psychogenic and functional interpretations of the illness presentation continue to be endorsed by some healthcare professionals, creating confusion and inappropriate diagnostic and therapeutic pathways for people living with long COVID. The purpose of this perspective is to present a clinical and scientific rationale for why long COVID should not be considered as a functional neurologic disorder. It will begin by discussing the parallel historical development of pathobiological and psychosomatic/sociogenic diagnostic constructs arising from a common root in neurasthenia, which has resulted in the collective understandings of myalgic encephalomyelitis/chronic fatigue syndrome (ME/CFS) and functional neurologic disorder (FND), respectively. We will also review the case definition criteria for FND and the distinguishing clinical and neuroimaging findings in FND vs. long COVID. We conclude that considering long COVID as FND is inappropriate based on differentiating pathophysiologic mechanisms and distinguishing clinical findings.

Keywords: post-COVID-19 condition (PCC); post-acute sequalae of COVID-19 (PASC); myalgic encephalomyelitis; chronic fatigue syndrome; neurasthenia; conversion disorder; dysautonomia; neurology; physical examination; imaging

1. Introduction

Severe fatigue that impairs usual function long has been described throughout recorded human history. The neurologists Beard [1] and Charcot [2] were among the first to characterize the health condition 'neurasthenia' in the latter half of the 19th century. Based on this common historical root in neurasthenia, two divergent scholarly and clinical paths have taken shape over time. The first path involves a pathogenic disease model rooted in the scientific process, resulting in a rich literature describing pathobiology and various attempts at creating specific case definition criteria. This path has resulted in the label of myalgic encephalomyelitis/chronic fatigue syndrome (ME/CFS). The second path is a psychosomatic/sociogenic illness construction that has incorporated ideas from contemporary neuroscience into an unbroken conceptual chain linking back to neurasthenia. This path has resulted in the label of functional neurologic disorders (FND).

Long COVID has caused a renewed scholarly and clinical focus on complex chronic conditions associated with infections [3–19]. According to the National Academy of Science, Engineering, and Medicine definition, long COVID is "an infection-associated chronic condition that occurs after SARS-CoV-2 infection and is present for at least 3 months as a continuous, relapsing and remitting, or progressive disease state that affects one or more organ systems" ([20], p. 2). Long COVID consists of single or multiple symptoms attributable to single or multiple diagnosable conditions [20]. It can follow asymptomatic, mild, or severe SARS-CoV-2 infection [20]. Among individuals with a positive COVID-19 test, approximately 43% of non-hospitalized cases and over half of hospitalized cases report symptoms and signs of long COVID, according to data from the first two years of the pandemic [21]. More recently, testing for SARS-CoV-2 infection has become far less frequent within society and surveillance testing has been discontinued. In addition, many individuals with COVID-19 now convalesce outside the medical system, so these cases are undetected. These observations suggest the prolonged decreases in function and delayed recovery [22–25] associated with COVID-19 may be undercounted and accelerating over time.

The intensity and disablement of fatigue associated with long COVID is similar to other post-viral conditions, including post-treatment Lyme disease [26], chronic Epstein–Barr infection [27], and post-mononucleosis syndrome [28]. The condition may range from mild impairment of function to severely disabling exhaustion. Patient complaints include severe waxing and waning fatigue, worsening fatigue the day after exertion, and dramatic exacerbation by efforts to exercise. Associated symptoms, including cognitive impairment (often referred to by patients as *brain fog*, diffuse chronic pain, sleep disruption, and autonomic dysfunction, including POTS, migraine, gastrointestinal dysmotility, and temperature intolerance are common concomitants. The onset of symptoms may be continuous from the time of infection or delayed in onset by weeks or months following an apparent full recovery from the acute phase of infection [20]. Long COVID disablement can range from mild to severe, and it can resolve in a period of months, or it can persist and worsen over time. Disablement related to long COVID may result in profound functional impairments in self-care, as well as family, social, school, and occupational roles [20]. Post-exertional malaise/post-exertional neuroimmune exhaustion (PEM/PENE) is common among people with long COVID [25,29,30], which accounts for the persistent, severe, and often progressive pattern of disablement in long COVID. PEM/PENE is a clinical hallmark of ME/CFS, suggesting an ME-like subtype of long COVID is prevalent [25,29,31–33]. Therefore, it is perhaps unsurprising that many of the same themes historically characterizing the narrative about ME/CFS are still influencing the discourse surrounding long COVID.

An accumulating body of research indicating the underlying pathophysiology of long COVID involves a complex interaction between processes and systems. Long COVID has been acknowledged to exacerbate pre-existing health conditions, or it may present as new diagnosable health conditions [20]. However, psychosomatic/sociogenic illness constructs continue to influence the contemporary discourse related to long COVID [34]. This clinical perspective will anchor the current discourse regarding long COVID into the historical context involving a parallel development of ME/CFS (predominately pathobiological) and FND (predominately psychosomatic/sociogenic) diagnostic constructs. This perspective will now review the clinical findings and neurobiological pathology of long COVID, developing a clinical and scientific rationale for why it is inappropriate to consider long COVID as FND.

2. Pathobiological Disease Characterization: From Neurasthenia to Myalgic Encephalomyelitis/Chronic Fatigue Syndrome

Some neurologists considered neurasthenia as a form of nervous exhaustion that caused severe mental and physical fatigue, even following the mild exertions associated with normal daily functions like self-care, family and community activities, and remunerative work [35]. Weakness characterized by abnormally rapid fatiguability and slow recovery following exertion were classically associated with neurasthenia. Yet, people with neurasthenia generally had unremarkable findings on physical examination despite having often severe functional limitations from a whole constellation of associated signs and symptoms. Nervous system exhaustion, nerve over-excitability, and impaired cerebral blood flow all were implicated as potential patho-etiological factors, perhaps secondary to overwork, toxicity, or infection [35]. The absence of remarkable physical findings consistent with neurasthenia, perhaps combined with a high prevalence of neurasthenia in women, led to early psychological theories suggesting that emotional disturbances corresponding to the severe physical and mental symptoms, signs, and disablement must be causal factors [35].

The personal and societal challenges of persistent fatigue never abated, even as the diagnosis of neurasthenia began to fall out of favor. It was during this time that the association between persistent fatigue and infection began to be more deeply explored. In 1934, Gilliam [36] documented an outbreak of infectious disease that caused lingering signs and symptoms at Los Angeles County General Hospital in the United States. Poliomyelitis was the best-known epidemic at the time, so Gilliam called this new condition atypical poliomyelitis [36]. Outbreaks of atypical poliomyelitis were also documented in Iceland in 1946–1947 and 1948–1949 [37]. In 1955, the term benign myalgic encephalomyelitis (ME) was introduced to describe the post-acute signs and symptoms following infectious disease at the Royal Free Hospital (London, UK) [38,39]. The term *epidemic ME* was then coined at a 1978 symposium of the Royal Society of Medicine [40]. This development was the medical community's first acknowledgement of ME as a distinct disease process, instead of a behavioral disorder. ME began to reach the popular consciousness in the mid-1980s US following an outbreak of post-infectious illness in Incline Village, Nevada. Work surrounding this outbreak led to assigning the name chronic fatigue syndrome (CFS) to signs and symptoms following an infection [4]. Although clinicians and researchers thought this term best described the phenomenon [41], people with ME believe it poorly represents their lived experience. Unsurprisingly, the term CFS remains deeply unpopular among people living with ME/CFS [42] even as it continues to find a common usage.

The nature of lingering symptoms, signs, and disability was the subject of exploration as the various outbreaks were documented. Ramsay first coined the term epidemic malaise to describe the phenomenon of muscle weakness that was worsened upon repeat testing [38,39]. This observation of a physical performance decline in response to a previous exertion was formative to developing contemporary case definition criteria for ME/CFS. PEM/PENE is now recognized as a whole host of unusual signs and symptoms following exertion, such as profound fatigue, cognitive dysfunction (such as impairment in attention, short-term memory, and performing mental calculations), sleep disturbance, clinical presentations consistent with viral reactivation (such as fevers, swollen glands, and pharyngitis), body and joint pains, headaches, and muscle weakness [43–50]. PEM/PENE appears responsible for the episodic disability observed in people living with ME/CFS. Episodic disability suggests a person's physical and cognitive abilities may vary substantially within a short term of hours to days (i.e., microcycling) and a long term of weeks, months, and years (i.e., macrocycling) [51,52]. In addition, PEM/PENE has increasingly become a component of case definition criteria over time to differentiate the phenomenon of debilitating fatigue, among other signs and symptoms, after exposure to a pathogen or toxin from other causes of fatigue.

Various case definitions to describe ME have been created throughout the late 20th century and early 21st century. These case definitions include the Holmes et al. [4] criteria (1988), Oxford criteria (1991) [53], Fukuda et al. criteria (1994) [53] and its elaboration by Reeves et al. (2005) [54], Canadian Consensus Criteria (CCC; 2003) [43], International Consensus Criteria for ME (ICC-ME; 2011) [44], criteria for Systemic Exertional Intolerance Disease (SEID; 2015) [55], and the UK National Institute for Health and Care Excellence guideline (UK NICE; 2021) [56]. There has been a progressively increasing prominence for the role of PEM/PENE as an important differentiating factor between ME/CFS and other health conditions associated with fatigue. PEM/PENE is now perhaps the most important specific (rule-in) consideration to identify ME/CFS and distinguish it from other health conditions that involve disabling fatigue.

ME was first discussed as being different from other neurological disease processes in a 1956 paper that first used the term "benign myalgic encephalitis" to distinguish it from other infectious encephalitic infections and, perhaps most importantly, hysteria [57]. It was first assigned an International Classification of Diseases (ICD) code in the ICD-8 1969 (code 332) [58]. ME and CFS are included in ICD-11 as post-viral syndromes (8E49) [59]. Inclusion of ME is evidence of improving legitimacy within the biomedical community, as the clinical characteristics and courses of these conditions have become better understood over time. Notably, ME and CFS are not included in the ICD as mental or behavioral disorders [59]. Key points in the development of a pathobiological disease construction resulting in the collective understanding of ME/CFS are summarized in Figure 1.

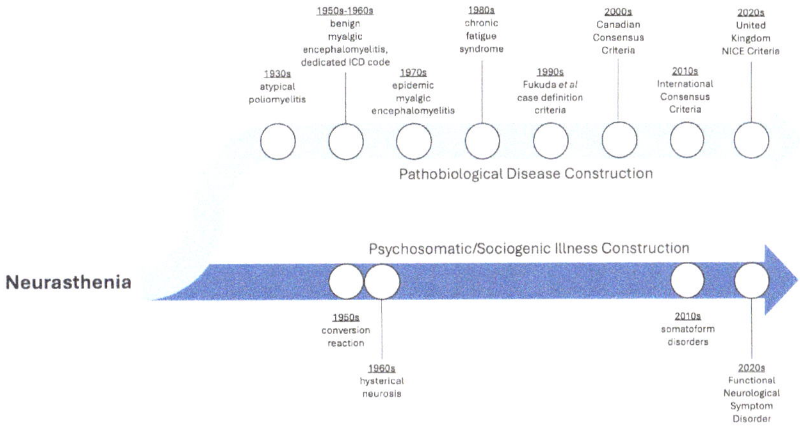

Figure 1. Key time points in the parallel development of disease and illness constructions resulting in myalgic encephalomyelitis/chronic fatigue syndrome (pathobiological illness construction) and functional neurologic disorder (psychosomatic/sociogenic illness construction), based on a common historical root in neurasthenia.

3. Psychosomatic/Sociogenic Illness Construction: From Neurasthenia to Functional Neurologic Disorder

While decades of scientific work have led down the path of iterative case definition criteria and the determining of the underlying pathophysiology of ME/CFS, a parallel path largely has repeated old thinking with new labels (Figure 1). The first edition of the Diagnostic and Statistical Manual (DSM) of the American Psychiatric Association listed hysteria as conversion reaction [60], transitioning to hysterical neurosis in the DSM's second edition [61]. These titles were based on the early concepts of hysteria as a uterine disorder in women [62]. A major underlying hypothesis advanced by Freud is that hysterical disorders involved the conversion between a somatic symptom and a repressed feeling or idea, such as a somatic symptom arising from anxiety, hence the term conversion disorder [63]. With the transition away from a system classifying disorders based on putative etiology and toward a contemporary system of psychodiagnostics by clinical phenomenology, the third edition of the DSM replaced hysterical neurosis with dissociative disorders and conversion disorders under the broader classification of somatoform disorders [62]. Early hypotheses regarding the etiology of hysteria were carried forward into thinking about somatoform disorders. Psychoanalytic theories suggested the repressed expression of conflicted unconscious drives, learning theories held that people with conversion disorders benefitted from secondary gain of their somatic symptoms, and sociocultural hypotheses were that somatic symptoms occur in substitution of the expression of intense forbidden ideas and emotions [63].

In 2013, the term 'functional neurological symptom disorder' was introduced in the DSM Version 5 Text Revision (DSM-V-TR) [64] and conversion disorder remained the main nomenclature. The 2022 revision of DSM-V-TR then changed the primary name to functional neurological symptom disorder and maintained conversion disorder as a synonym (Box 1) [65]. FND is now classified by the International Classification of Diseases (ICD-11) [59] as a dissociative neurological symptom disorder, defined as a mental health condition involving a loss of connection between thoughts, memories, feelings, surroundings, behavior, and identity [66]. More recent data from neuroscientific studies [67–72] have been used to support claims of emotional processing that might be familiar to earlier advocates of hysteria and somatoform conditions. Despite poor-quality supporting research [73], mainstay interventions for FND continue to include psychodynamic and cognitive-behavior therapies to address emotional processing. Thus, despite the original intent [74] and subsequent rationalizations [75] of the principal proponents of FND, this brief historical analysis indicates a continuous underlying conceptual thread that remains unbroken between neurasthenia, through hysterical neurosis and somatoform disorders, leading to the contemporary psychosomatic/sociogenic illness construction of FND.

Box 1. Diagnostic criteria for functional neurologic disorder [61]

- One or more symptoms of altered voluntary motor behavior or sensory function
- Clinical findings provide evidence of incompatibility between the symptom and recognized neurological or medical conditions
- The symptom or deficit is not better explained by another medical or mental disorder
- The symptom or deficit causes clinically significant distress or impairment in social, occupational, or other important areas of functioning, or warrants medical evaluation

Proponents suggest that just the label of FND may be helpful for some patients who live with troublesome but medically unexplained symptoms and signs [76,77]. Even still, the DSM-V-TR diagnostic criteria for FND indicate this label should not be provided when an alternative diagnosis is more compelling. For example, ME/CFS should not be considered as a functional disorder because more specific case definition criteria best explain its constellation of symptoms, signs, and pathophysiology. However, some prominent medical organizations have conflated PEM/PENE with FND because some symptoms of PEM/PENE are represented among the DSM-V-TR case definition criteria for FND. For example, the UK NICE attempted to classify ME/CFS as a functional disorder in 2017 [78]. This action was met with significant opposition from the ME community [79]. The dispute lasted over two years with the subsequent guideline removing the reference to ME/CFS as FND [78]. Both individual clinicians (generally from the fields of neurology and psychiatry) and prominent national medical organizations in European countries [80,81] persist in classifying ME/CFS as FND despite compelling evidence to the contrary.

4. Evidence Refutes That Long COVID Should Be Considered a Functional Neurologic Disorder

FND refers to medical and neurologic symptoms that fail to match any existing medical or neurological conditions [82]. It is a rare syndrome, affecting around 4–12 per 100,000 people despite a suggestion that it is commonly diagnosed in neurology clinics [83]. FND is usually diagnosed when patients are observed to experience seizure-like spells in the setting of normal electroencephalography (EEG) or when they demonstrate abnormal movements or paralysis that are incongruent with their neurologic exam and neuroimaging. In some patients, FND may occur alongside other diagnosable entities, such as long COVID.

In this scenario, the clinician should evaluate and treat those diagnosable entities, and refrain from considering the entire patient presentation as "functional" simply because functional aspects may be present. While the rate of misdiagnosis with FND in patients with long COVID is unknown, clinical experience suggests that many patients with complex chronic disorders in general have been misdiagnosed with anxiety, depression, or FND at some point in the course of their illness (Box 2).

Box 2. Defined conditions commonly misdiagnosed as functional neurologic disorder

- **Autonomic conditions**, such as neurocardiogenic syncope, postural orthostatic tachycardia syndrome, orthostatic intolerance, and autonomic and small fiber neuropathy
- **Chronic pain conditions**, such as fibromyalgia, myofascial pain syndrome, and complex regional pain syndrome
- **Systemic immune conditions**, such as mast cell activation syndrome and mastocytosis
- **Autoimmune conditions**, such as Sjögren's syndrome, systemic lupus erythematosus, and anti-phospholipid syndrome
- **Genetic conditions**, such as hypermobile Ehlers-Danlos syndrome and other hypermobility spectrum disorders, Fabry's disease and others
- **Mitochondrial and metabolic conditions**
- **Infection-associated chronic conditions**, such as myalgic encephalomyelitis, Long Covid/Post-Covid condition, and post-treatment Lyme disease

4.1. Refutative Evidence from Pathophysiology

The scope and severity of Long COVID-related disablement in individuals and in society has incentivized investigations into the pathophysiology of this novel infection-associated chronic disease. Long COVID is now understood as an umbrella term encompassing a complex pathophysiology affecting multiple organ systems. Various potential aspects of long COVID pathobiology include autonomic manifestations [84]; vascular and endothelial dysfunction in the context of hypercoagulability [85–87]; viral persistence [88]; abnormalities in T cell populations and responses [89,90]; impaired cardiopulmonary function [91,92]; autoimmunity [93–95]; bioenergetic impairments [96–98]; small fiber neuropathy [99]; and alterations in the gut microbiome [100]. In 2021, post-COVID-19 condition (or, long COVID) was assigned an ICD code (U09.9) [101]. The collective understanding of long COVID is far from settled. However, an accumulating science now provides a more compelling pathophysiological basis for testing and interventions than considering long COVID as a functional disorder. Long COVID should not be broadly considered as FND because of its "organic" nature, requiring FND to be ruled out according to DSM-V-TR criteria [64].

4.2. Refutative Evidence from Clinical Presentation

Typically, people with functional disorders often exhibit numerous multi-systemic and multi-organ concerns, including various neurologic and psychiatric manifestations such as sensory disturbance, motor weakness, balance difficulty, chronic dizziness, chronic vertigo, chronic pain, chronic fatigue, sleep impairment, urinary and gastrointestinal symptoms, and cognitive dysfunction (Table 1). The symptom experience and distress associated with symptoms in people with FND is frequently not supported or incongruent with objective findings on neurologic examination and diagnostic testing. People with functional disorders also may have comorbid psychiatric conditions, such as depression and anxiety [102–105]. It remains unclear whether the prevalence and severity of psychiatric comorbidities among those with functional disorders is greater than people living with other types of chronic illnesses, and whether psychiatric conditions contributed to other signs and symptoms or are a secondary reaction to their presence. In addition, inventories used to measure anxiety and depression often may capture the autonomic signs and symptoms of underlying pathobiological process. FND, if diagnosed correctly through positive signs on neurologic examination, does not appear to be as common although true prevalence is unknown and needs to be studied. Moreover, no carefully designed research studies have been conducted to systematically test the hypothesis that long COVID is a functional disorder and the extant literature is poor in methodological quality [106].

Table 1. Typical clinical features of long COVID, myalgic encephalomyelitis/chronic fatigue syndrome, and functional neurologic disorder.

Clinical Feature	ME/CFS	Long COVID	FND
Post-exertional malaise/ Post-exertional neuroimmune exhaustion	Yes	Yes, some types	No
Pain	Yes	Yes, some types	Sometimes
Dizziness	Yes	Yes	No
Neuropathic features	Yes	Yes	No
Recurrent flu-like symptoms	Yes	Common	No
Dysautonomia	Yes	Common	No
Abnormal sleep study	Yes	Yes	No
Fatigue	Yes	Yes	Yes
Impaired sleep	Yes	Yes	Yes
Functional leg weakness	No	No	Yes
Functional seizures	No	No	Yes
Functional tremor	No	No	Yes
Functional dystonia	No	No	Yes
Functional gait disorder	No	No	Yes
Functional facial spasm	No	No	Yes
Functional tics	No	No	Yes
Functional drop attacks	No	No	Yes
Functional sensory symptoms	No	No	Yes
Functional cognitive symptoms	No	No	Yes
Functional speech and swallowing	No	No	Yes
Functional visual symptoms	No	No	Yes
Dissociative symptoms	No	No	Yes

4.2.1. Motor Examination

The diagnosis of FND requires the presence of discrete neurologic deficits, which are usually elicited as part of the neurologic examination (Tables 1 and 2). Presenting features may include weakness in the lower or upper extremities of sudden onset and can be unilateral or bilateral. A neurologic examination is used to demonstrate evidence of internal inconsistency between voluntary movements and automatic movements through findings of a positive Hoover's sign and hip abductor sign [107]. While weakness in the extremities is a common concern in many people living with long COVID, clinical experience indicates that the neurologic examination typically reveals an unremarkable motor examination without Hoover's or hip abductor signs. Give-way weakness may be present in patients with long COVID, but it is usually diffuse and non-lateralizing and occurs secondary to pain, fatigue, PEM/PENE, or orthostatic intolerance. In the context of these companion findings, give-way weakness should not be interpreted as evidence of a functional etiology.

Table 2. Key differentiating physical examination findings of long COVID vs. functional neurologic disorder.

Findings	Long COVID	Functional Neurological Disorder
Vital Signs	• Postural tachycardia • Orthostatic hypotension • Dizziness • Other symptoms upon standing that are relieved by sitting or lying down	• No usual abnormalities
Cranial Nerves	• May have dilated poorly reactive pupils or mild horizontal end-point nystagmus	• Normal, although patients may report vision or hearing impairment
Motor	• Give-way weakness may be present due to fatigue, post-exertional neuroimmune exhaustion, misunderstanding the task, or poor effort	• Weakness inconsistent with known neurologic patterns • Paralysis and weakness with positive Hoover's sign and/or positive hip abductor sign
Sensory	• Length and non-length dependent reduced temperature and pinprick consistent with small fiber neuropathy	• Complete anesthesia in certain body parts or exactly at the midline or below the waist, incongruent with known neurologic patterns
Movement	• Whole-body shaking, mild postural tremor, and/or internal tremor not visible to the examiner • Possible fasciculations due to benign fasciculation syndrome	• Tremor entrainment • Tremor that disappears with distraction • Inconsistent tremor • Unusual tremor incongruent with neurologic disorders
Gait	• Usually normal, but some unsteadiness and difficulty with tandem walking might be present	• Functional gait • Astasia-abasia • Unusual gait pattern inconsistent with another neurological cause
Observation	• Syncope • Presyncope • Orthostatic intolerance • Anoxic seizures	• Spells with non-epileptic convulsions
Skin	• Acrocyanosis with discoloration of the legs and/or arms distally, more in the dependent position (Figure 2) • Possible dermatographia • Possible dry skin • Possible pale or flushed appearance • Possible maculopapular rashes, urticarial lesions, and chilblains	• No usual abnormalities

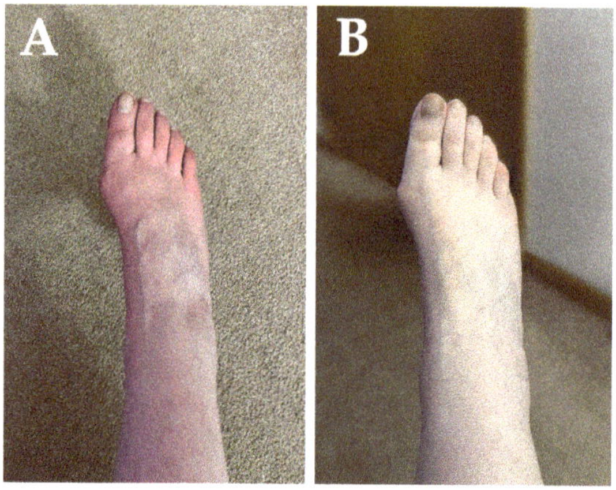

Figure 2. Acrocyanosis in the distal leg in the dependent position (**A**) that immediately disappears on raising the leg up against gravity (**B**) in a person with long COVID. She was initially misdiagnosed with functional neurologic disorder based on non-epileptic spells, which were subsequently determined to be pre-syncopal episodes caused by post-COVID-19 postural orthostatic tachycardia syndrome.

4.2.2. Sensory Examination

A common presenting feature of FND is sensory disturbance that fails to fit into defined patterns of neuropathy, radiculopathy, or the central lesion of the brain (Tables 1 and 2). Sensory testing as part of neurologic examination in a patient with FND may reveal complete anesthesia in non-anatomic distributions, such as involving an entire extremity, located exactly at the midline, or below the waist. While sensory disturbance is common in patients with long COVID due, in part, to post-COVID-19 small fiber neuropathy [108], sensory exam findings usually correspond to a neuropathic pattern with decreased pinprick and temperature sensations in the feet or hands, distally more than proximally. However, a patchy sensory loss distribution is not uncommon in those who present with non-length-dependent patchy small fiber neuropathy, a form that is especially prevalent in people with autoimmune disorders [109].

4.2.3. Tremor

Tremor may be another presenting feature of FND with examination findings revealing tremor entrainment and other inconsistent tremor characteristics (Tables 1 and 2). Tremor and other abnormal movements may be among common complaints of patients with long COVID, but often involve diffuse, whole-body body tremors or shaking, which may be associated with dysautonomia and hyperadrenergic state, including abnormal blood pressure, heart rate, and blood volume. Autonomic dysfunction affects nearly 70% of patients with long COVID [110], so improvement or resolution of abnormal movements associated with dysautonomia may be noted with hydration, increased salt intake, or medications. Additionally, the sensation of "internal vibrations" is often described by people living with long COVID; although the etiology of this concern is not fully understood, clinical experience suggests it often occurs in patients with hypovolemia, dysautonomia, and small fiber neuropathy. These features should not be attributed to functional causes or FND.

4.2.4. Spells and Seizures

Spells of unknown etiology are often attributed to FND, especially when accompanied by normal EEG in the setting of convulsive activity (Table 1). While a small subset of people living with long COVID could have non-epileptic functional seizures, many patients with long COVID have post-COVID-19 dysautonomia in the form of neurocardiogenic syncope, postural orthostatic tachycardia syndrome, orthostatic hypotension, and orthostatic intolerance. These patients often experience spells of presyncope or syncope, some with convulsive activity during syncope, which is termed anoxic seizures. Patients with presyncope and syncope may be misdiagnosed with FND by neurologists with limited knowledge of the phenomenology of syncope. In cases where the etiology of spells is unclear, a tilt table test can provide differentiation between syncope and pseudo-syncope with adequate sensitivity and specificity [111]. A video recording of the spells obtained by the family also may be reviewed by a neurologist to assist with differentiating between syncope and dissociative/functional seizures.

4.2.5. Gait Examination

Gait examination of people living with long COVID is typically unremarkable, although some people may have unsteadiness due to orthostasis or poor proprioception related to large fiber neuropathy, chronic vestibulopathy, or hypermobility spectrum disorders (Tables 1 and 2). Clinical features of abnormal gait and movements consistent with FND, such as dystonia, ticks, twitches, and jerks, are typically uncommon in people with long COVID. Concerns about gait and findings in the gait examination explainable by other causes should not be taken as signs of FND.

4.2.6. Urinary Functioning

Urinary retention is sometimes listed as a feature of FND [112], but clinical experience suggests that urinary retention rarely occurs in people with long COVID. Urinary retention is a common feature of autonomic neuropathy and, if present in patients with long COVID, should prompt an investigation for post-COVID-19 autonomic neuropathy or ganglionopathy which have been described as rare post-COVID-19 conditions [113].

4.2.7. Cognition

Cognitive concerns, such as difficulty with attention, concentration, and memory, have been endorsed by some proponents of FND as being functional in nature. One review suggested that almost one quarter of patients attending memory clinics may have functional cognitive disorders [114]. However, it is unclear how these complaints are differentiated from the cognitive impairment—commonly referred to as brain fog—in patients with long COVID. Numerous studies suggested neuroinflammation and microglial activation as mechanisms of post-COVID-19 neurologic sequelae [115,116]. Importantly, several studies identified neuropsychological deficits via cognitive testing in patients with long COVID [117,118].

Moreover, traditional screening tests for evaluation of cognitive impairment designed to screen patients for Alzheimer's disease are not useful in patients with cognitive impairment secondary to long COVID [119]. Thus, currently available screening tests that were designed for neurodegenerative conditions and not neuroinflammatory or neuroimmune conditions are inadequate to rule in or out impairments secondary to neuroinflammatory and neuroimmune processes, and, therefore, cannot be utilized to diagnose as functional by default if results are "normal" in patients with long COVID and other disorders that are associated with cognitive complaints. Cognitive tests to assess patients with non-neurodegenerative cognitive complaints need to be designed to provide clinicians with validated tools to better evaluate and quantify the extent of cognitive impairment in patients with post-COVID-19 neurocognitive syndrome.

4.2.8. Summary

In summary, FND and long COVID can be effectively differentiated through a comprehensive clinical examination. One caveat is the difficulty that arises when a patient presents with some evidence of functional neurologic disorder on physical examination (e.g., with tremor entrainment or positive Hoover's sign) in conjunction with postural tachycardia, acrocyanosis, and other features of long COVID and post-COVID-19 dysautonomia or small fiber neuropathy (Figure 2). In cases like these, management of non-FND disorders and symptoms should be the top priority. Clinical experience suggests that a significant number of patients with long COVID are being misdiagnosed with FND without diagnosing and addressing the primary long COVID pathophysiology or symptoms, such as dizziness, palpitations, tachycardia, and pain. In these cases, FND-tailored diagnostic and therapeutic approaches delay improvement and recovery by failing to implement the pharmacologic and non-pharmacologic therapies targeting underlying autonomic, neuropathic, and cardiovascular pathophysiologies. Referral to FND-tailored rehabilitation programs should be considered only for patients, in whom post-COVID-19 FND is determined to be the main component of long COVID and in strict adherence with relevant case definition criteria [82].

4.3. *Refutative Evidence from Neuroimaging*

Despite its relatively recent recognition, the literature describing significant structural brain abnormalities in long COVID is already extensive. This literature suggests FND cannot commonly explain long COVID, because it indicates neurologic signs, symptoms, and disability may be caused by structural changes in the brain. Such abnormalities have been demonstrated using various imaging modalities and range from changes in gray matter thickness and volume, to macro- and microstructural white matter changes and evidence of metabolic and neuroinflammatory derangement. It is also important to note that despite clearly distinctive systemic immunological abnormalities [120–122] and abnormal neuroimmune profiles [121,123] in long COVID, routine clinical structural magnetic resonance imaging (MRI) sequences often return normal findings [123].

Douaud et al. [124] examined structural brain changes in a biobank cohort based in the United Kingdom, before and after SARS-CoV-2 infection. Compared to uninfected controls, the authors found significant gray matter thickness reduction following infection, with a reduction in global brain size. Hosp et al. [125] used an MRI diffusion microstructure imaging technique to evaluate subtle changes in both gray and white matter integrity. Compared to recovered infected patients, those with ongoing symptoms demonstrated widespread changes in microstructure, which correlated with evaluations of cognitive dysfunction. Wu et al. [126] used another diffusion tensor imaging (DTI) technique to evaluate the perivascular space and glymphatic system. These authors calculated flow in the perivascular spaces alongside medullary veins, which lie orthogonal to the projection and association nerve fibers in the periventricular deep white matter. They reported reduction in the indices for glymphatic function in people living with long COVID even following a mild acute infection. Another small cohort study compared recovered, brain fog positive, and brain fog negative patients with long COVID [127]. Dynamic contrast-enhanced MRI (DCE-MRI) showed significant whole brain leakage, indicating increased blood–brain barrier (BBB) permeability, in only the 'brain fog' sub-group [127]. Chaganti et al. combined techniques in a longitudinal study of 14 patients with long COVID-related cognitive impairment [128]. DCE-MRI and DTI revealed impairments in the integrity of BBB and white matter microstructure [128]. Simultaneously, MR spectroscopy demonstrated reduced glutamate/glutamine in these areas, leading the authors to suggest that white matter injury may result from glutamatergic excitotoxicity, secondary to reduced BBB integrity associated with neuroinflammation [128]. VanElzakker et al. [129] used positron emission tomography (PET) with a tracer for activated microglia ([11C]PBR28) to report evidence of significantly increased neuroinflammation in many brain regions in LC. Peluso et al. [130] used a novel PET tracer ([18F]F-AraG) to tag activated T cells. Following SARS-CoV-2 infection, activated T cells were found in multiple organs including the bowel and bone marrow, but notably had trafficked into central nervous system (CNS) sites such as the brainstem and spinal cord, where they should be absent. This finding was more exaggerated in patients with long COVID signs and symptoms [130]. Biopsy-accessible tissues such as colon tissue demonstrated residual viral components, and the authors speculated they also might be present in the CNS [130].

A 2021 review article canvased the literature of neuroimaging in FND [131]. The highlighted modalities were functional MRI (fMRI), using both resting-state and task-based paradigms; high-resolution structural MRI evaluation of gray matter; DTI of white matter microstructure; MR spectroscopy; CT/MR positron emission tomography; and near-infrared spectroscopy. The authors note that neuroimaging in FND is early in its development, with few replicated studies, and with confounding factors in terms of clinical heterogeneity and co-morbidities. They conclude by encouraging a multimodal neuroimaging approach to advance the field. Most fMRI studies using blood oxygen level-dependent (BOLD) techniques have shown abnormalities in specific brain regions, yet data have been inconsistent [132]. Very recently, Schneider et al. [133] have attempted to further define the variability of BOLD signal in FND, with particular emphasis on the somatomotor, limbic, and salience networks. However, when structural abnormalities have been found in gray [134] or white matter [67], it remains unclear whether they are a cause, consequence, or comorbidity [132,135].

While neuroimaging in FND is an evolving field, there are already replicated findings in long COVID that point toward a coherent structural pathophysiology. Aspects highlighted in the literature to date involve neuroinflammation with microglial activation, a dysfunctional blood–brain barrier, white matter microstructural changes, as well as reduction in gray matter volume. Systemic dysfunction, such as orthostatic intolerance with reduced cerebral blood flow, is also shown to be a key contributor to symptoms [136–140]. The detail of how these findings are driven from specific and potentially correctable upstream causes is enthusiastically anticipated by patients, clinicians, and researchers alike.

5. Conclusions

Long COVID continues to be a major public health issue [141]. While several phenotypes of long COVID clinical presentation have emerged based on observational studies and collective clinical experience over the past four years, it is important to emphasize that the vast majority of patients with long COVID do not have FND. As this perspective indicates, long COVID is not based in 'functional' etiology, as demonstrated by numerous studies identifying a complex pathophysiology as well as common findings from the clinical examination and a summary of extant structural neuroimaging studies. Further research is needed to delineate precise pathophysiological pathways and effective therapies for long COVID and numerous post-COVID-19 neurologic manifestations. Additionally, studies applying accepted case definition criteria are also needed to determine the true prevalence of FND as the sole or major contributor to symptoms and disability among individuals with persistent symptoms following SARS-CoV-2 infection. These studies will help establish clinical practices that best differentiate this small subset of patients with FND-related specialized needs from the vast majority of people who experience long COVID in the forms of ME/CFS, dysautonomia, immune dysfunction, small fiber neuropathy and other post-COVID-19 neurologic syndromes.

Author Contributions: Conceptualization, T.E.D., S.B., D.D.-P., N.C.-B., G.J.T. and S.F.T.; writing-original draft preparation, T.E.D., S.B., D.D.-P., N.C.-B., G.J.T. and S.F.T.; writing-review and editing, T.E.D., S.B., D.D.-P. and N.C.-B. All authors have read and agreed to the published version of the manuscript.

Funding: This research received no external funding.

Institutional Review Board Statement: Not applicable.

Informed Consent Statement: Not applicable.

Data Availability Statement: Not applicable.

Conflicts of Interest: The authors declare no conflicts of interest.

References

1. Beard, G. Neurasthenia, or nervous exhaustion. *Boston Med. Surg. J.* **1869**, *3*, 217–221. [CrossRef]
2. Goetz, C.G. Poor Beard!! Charcot's internationalization of neurasthenia, the "American disease". *Neurology* **2001**, *57*, 510–514. [CrossRef]
3. Hickie, I.; Davenport, T.; Wakefield, D.; Vollmer-Conna, U.; Cameron, B.; Vernon, S.D.; Reeves, W.C.; Lloyd, A.; Dubbo Infection Outcomes Study Group. Post-infective and chronic fatigue syndromes precipitated by viral and non-viral pathogens: Prospective cohort study. *BMJ* **2006**, *333*, 575. [CrossRef]
4. Holmes, G.P.; Kaplan, J.E.; Gantz, N.M.; Komaroff, A.L.; Schonberger, L.B.; Straus, S.E.; Jones, J.F.; Dubois, R.E.; Cunningham-Rundles, C.; Pahwa, S.; et al. Chronic fatigue syndrome: A working case definition. *Ann. Intern. Med.* **1988**, *108*, 387–389. [CrossRef]
5. Ikuta, K.; Yamada, T.; Shimomura, T.; Kuratsune, H.; Kawahara, R.; Ikawa, S.; Ohnishi, E.; Sokawa, Y.; Fukushi, H.; Hirai, K.; et al. Diagnostic evaluation of 2′, 5′-oligoadenylate synthetase activities and antibodies against Epstein-Barr virus and Coxiella burnetii in patients with chronic fatigue syndrome in Japan. *Microbes Infect.* **2003**, *5*, 1096–1102. [CrossRef]
6. Klimas, N.G.; Salvato, F.R.; Morgan, R.; Fletcher, M.A. Immunologic abnormalities in chronic fatigue syndrome. *J. Clin. Microbiol.* **1990**, *28*, 1403–1410. [CrossRef]
7. Kogelnik, A.M.; Loomis, K.; Hoegh-Petersen, M.; Rosso, F.; Hischier, C.; Montoya, J.G. Use of valganciclovir in patients with elevated antibody titers against Human Herpesvirus-6 (HHV-6) and Epstein-Barr Virus (EBV) who were experiencing central nervous system dysfunction including long-standing fatigue. *J. Clin. Virol.* **2006**, *37* (Suppl. S1), S33–S38. [CrossRef]
8. Lerner, A.M.; Beqaj, S.H.; Deeter, R.G.; Fitzgerald, J.T. IgM serum antibodies to Epstein-Barr virus are uniquely present in a subset of patients with the chronic fatigue syndrome. *In Vivo* **2004**, *18*, 101–106.
9. Shikova, E.; Reshkova, V.; Kumanova, A.; Raleva, S.; Alexandrova, D.; Capo, N.; Murovska, M.; On Behalf of the European Network on Me/Cfs Euromene. Cytomegalovirus, Epstein-Barr virus, and human herpesvirus-6 infections in patients with myalgic encephalomyelitis/chronic fatigue syndrome. *J. Med. Virol.* **2020**, *92*, 3682–3688. [CrossRef]
10. Kasimir, F.; Toomey, D.; Liu, Z.; Kaiping, A.C.; Ariza, M.E.; Prusty, B.K. Tissue specific signature of HHV-6 infection in ME/CFS. *Front. Mol. Biosci.* **2022**, *9*, 1044964. [CrossRef]
11. Sejvar, J.J.; Curns, A.T.; Welburg, L.; Jones, J.F.; Lundgren, L.M.; Capuron, L.; Pape, J.; Reeves, W.C.; Campbel, G.L. Neurocognitive and functional outcomes in persons recovering from West Nile virus illness. *J. Neuropsychol.* **2008**, *2*, 477–499. [CrossRef]
12. Prevail III Study Group; Sneller, M.C.; Reilly, C.; Badio, M.; Bishop, R.J.; Eghrari, A.O.; Moses, S.J.; Johnson, K.L.; Gayedyu-Dennis, D.; Hensley, L.E.; et al. A longitudinal study of Ebola sequelae in Liberia. *N. Engl. J. Med.* **2019**, *380*, 924–934. [CrossRef]
13. Kelly, J.D.; Van Ryn, C.; Badio, M.; Fayiah, T.; Johnson, K.; Gayedyu-Dennis, D.; Weiser, S.D.; Porco, T.C.; Martin, J.N.; Sneller, M.C.; et al. Clinical sequelae among individuals with pauci-symptomatic or asymptomatic Ebola virus infection and unrecognised Ebola virus disease in Liberia: A longitudinal cohort study. *Lancet Infect. Dis.* **2022**, *22*, 1163–1171. [CrossRef]
14. Chia, J.K. The role of enterovirus in chronic fatigue syndrome. *J. Clin. Pathol.* **2005**, *58*, 1126–1132. [CrossRef]
15. Chia, J.K.; Chia, A.Y. Chronic fatigue syndrome is associated with chronic enterovirus infection of the stomach. *J. Clin. Pathol.* **2008**, *61*, 43–48. [CrossRef]
16. Ledina, D.; Bradaric, N.; Milas, I.; Ivic, I.; Brncic, N.; Kuzmicic, N. Chronic fatigue syndrome after Q fever. *Med. Sci. Monit.* **2007**, *13*, CS88–CS92.
17. Treib, J.; Grauer, M.T.; Haass, A.; Langenbach, J.; Holzer, G.; Woessner, R. Chronic fatigue syndrome in patients with Lyme borreliosis. *Eur. Neurol.* **2000**, *43*, 107–109. [CrossRef]
18. Fares-Medina, S.; Diaz-Caro, I.; Garcia-Montes, R.; Corral-Liria, I.; Garcia-Gomez-Heras, S. Multiple Chemical Sensitivity Syndrome: First symptoms and evolution of the clinical picture: Case-control study/epidemiological case-control study. *Int. J. Environ. Res. Public Health* **2022**, *19*, 15891. [CrossRef]
19. Katerndahl, D.A.; Bell, I.R.; Palmer, R.F.; Miller, C.S. Chemical intolerance in primary care settings: Prevalence, comorbidity, and outcomes. *Ann. Fam. Med.* **2012**, *10*, 357–365. [CrossRef]
20. National Academies of Sciences, Engineering, and Medicine. *A Long COVID Definition: A Chronic, Systemic Disease State with Profound Consequences*; The National Academies Press: Washington, DC, USA, 2024.
21. Chen, C.; Haupert, S.R.; Zimmermann, L.; Shi, X.; Fritsche, L.G.; Mukherjee, B. Global Prevalence of Post COVID-19 Condition or long COVID: A meta-analysis and systematic review. *J. Infect. Dis.* **2022**, *226*, 1593–1607. [CrossRef]
22. Xie, Y.; Xu, E.; Bowe, B.; Al-Aly, Z. Long-term cardiovascular outcomes of COVID-19. *Nat. Med.* **2022**, *28*, 583–590. [CrossRef] [PubMed]
23. Bowe, B.; Xie, Y.; Al-Aly, Z. Postacute sequelae of COVID-19 at 2 years. *Nat. Med.* **2023**, *29*, 2347–2357. [CrossRef] [PubMed]
24. Cai, M.; Xie, Y.; Topol, E.J.; Al-Aly, Z. Three-year outcomes of post-acute sequelae of COVID-19. *Nat. Med.* **2024**, *30*, 1564–1573. [CrossRef] [PubMed]
25. Davis, H.E.; Assaf, G.S.; McCorkell, L.; Wei, H.; Low, R.J.; Re'em, Y.; Redfield, S.; Austin, J.P.; Akrami, A. Characterizing long COVID in an international cohort: 7 months of symptoms and their impact. *eClinicalMedicine* **2021**, *38*, 101019. [CrossRef] [PubMed]
26. Bai, N.A.; Richardson, C.S. Posttreatment Lyme disease syndrome and myalgic encephalomyelitis/chronic fatigue syndrome: A systematic review and comparison of pathogenesis. *Chronic Dis. Transl. Med.* **2023**, *9*, 183–190. [CrossRef] [PubMed]

27. Ruiz-Pablos, M.; Paiva, B.; Montero-Mateo, R.; Garcia, N.; Zabaleta, A. Epstein-Barr Virus and the origin of myalgic encephalomyelitis or chronic fatigue syndrome. *Front. Immunol.* **2021**, *12*, 656797. [CrossRef] [PubMed]
28. Jason, L.A.; Katz, B.Z.; Shiraishi, Y.; Mears, C.J.; Im, Y.; Taylor, R. Predictors of post-infectious chronic fatigue syndrome in adolescents. *Health Psychol. Behav. Med.* **2014**, *2*, 41–51. [CrossRef] [PubMed]
29. Jason, L.A.; Dorri, J.A. ME/CFS and post-exertional malaise among patients with long COVID. *Neurol. Int.* **2022**, *15*, 1–11. [CrossRef] [PubMed]
30. Twomey, R.; DeMars, J.; Franklin, K.; Culos-Reed, S.N.; Weatherald, J.; Wrightson, J.G. Chronic fatigue and postexertional malaise in people living with long COVID: An observational study. *Phys. Ther.* **2022**, *102*, pzac005. [CrossRef] [PubMed]
31. Vernon, S.D.; Hartle, M.; Sullivan, K.; Bell, J.; Abbaszadeh, S.; Unutmaz, D.; Bateman, L. Post-exertional malaise among people with long COVID compared to myalgic encephalomyelitis/chronic fatigue syndrome (ME/CFS). *Work* **2023**, *74*, 1179–1186. [CrossRef] [PubMed]
32. Jason, L.A.; Islam, M.; Conroy, K.; Cotler, J.; Torres, C.; Johnson, M.; Mabie, B. COVID-19 symptoms over time: Comparing long-haulers to ME/CFS. *Fatigue* **2021**, *9*, 59–68. [CrossRef] [PubMed]
33. Tokumasu, K.; Honda, H.; Sunada, N.; Sakurada, Y.; Matsuda, Y.; Yamamoto, K.; Nakano, Y.; Hasegawa, T.; Yamamoto, Y.; Otsuka, Y.; et al. Clinical characteristics of myalgic encephalomyelitis/chronic fatigue syndrome (ME/CFS) diagnosed in patients with long COVID. *Medicina* **2022**, *58*, 850. [CrossRef] [PubMed]
34. Little, J.; Higgins, M.; Palepu, R. Long COVID—Can we deny a diagnosis without denying a person's reality? *Australas. Psychiatry* **2024**, *32*, 44–46. [CrossRef] [PubMed]
35. Leitch, A.G. Neurasthenia, myalgic encephalitis or cryptogenic chronic fatigue syndrome? *QJM* **1995**, *88*, 447–450.
36. Gilliam, A.G. *Epidemiological Study on an Epidemic, Diagnosed as Poliomyelitis, Occurring among the Personnel of Los Angeles County General Hospital during the Summer of 1934*; United States Government Printing Office: Washington, DC, USA, 1934.
37. Sigurdsson, B.; Sigurjonsson, J.; Sigurdsson, J.H.; Thorkelsson, J.; Gudmundsson, K.R. A disease epidemic in Iceland simulating poliomyelitis. *Am. J. Hyg.* **1950**, *52*, 222–238. [CrossRef] [PubMed]
38. Ramsay, A.M. Myalgic encephalomyelitis: A baffling syndrome with a tragic aftermath. *ME Assoc. J.* **1981**, *153*, 40–41.
39. Ramsay, A.M. *Myalgic Encephalomyelitis and Postviral Fatigue States: The Saga of Royal Free Disease*; Gower Medical Publishing: London, UK, 1986.
40. Epidemic myalgic encephalomyelitis. *Br. Med. J.* **1978**, *1*, 1436–1437. [CrossRef]
41. Evengard, B.; Schacterle, R.S.; Komaroff, A.L. Chronic fatigue syndrome: New insights and old ignorance. *J. Intern. Med.* **1999**, *246*, 455–469. [CrossRef]
42. Jason, L.A.; Eisele, H.; Taylor, R.R. Assessing attitudes toward new names for chronic fatigue syndrome. *Eval. Health Prof.* **2001**, *24*, 424–435. [CrossRef] [PubMed]
43. Carruthers, B.M.; Jain, A.K.; DeMeirleir, K.L.; Peterson, D.L.; Klimas, N.G.; Lerner, A.M.; Flor-Henry, P.; Joshi, P.; Powles, A.C.; Sherkey, J.A.; et al. Myalgic encephalomyelitis/chronic fatigue syndrome: Clinical working case definition, diagnostic and treatment protocols. *J. Chronic Fatigue Syndr.* **2003**, *1*, 7–115. [CrossRef]
44. Carruthers, B.M.; van de Sande, M.I.; De Meirleir, K.L.; Klimas, N.G.; Broderick, G.; Mitchell, T.; Staines, D.; Powles, A.C.; Speight, N.; Vallings, R.; et al. Myalgic encephalomyelitis: International Consensus Criteria. *J. Intern. Med.* **2011**, *270*, 327–338. [CrossRef] [PubMed]
45. Chu, L.; Valencia, I.J.; Garvert, D.W.; Montoya, J.G. Deconstructing post-exertional malaise in myalgic encephalomyelitis/chronic fatigue syndrome: A patient-centered, cross-sectional survey. *PLoS ONE* **2018**, *13*, e0197811. [CrossRef] [PubMed]
46. Chu, L.; Valencia, I.J.; Garvert, D.W.; Montoya, J.G. Onset patterns and course of myalgic encephalomyelitis/chronic fatigue syndrome. *Front. Pediatr.* **2019**, *7*, 12. [CrossRef] [PubMed]
47. Fukuda, K.; Straus, S.E.; Hickie, I.; Sharpe, M.C.; Dobbins, J.G.; Komaroff, A. The chronic fatigue syndrome: A comprehensive approach to its definition and study. International Chronic Fatigue Syndrome Study Group. *Ann. Intern. Med.* **1994**, *121*, 953–959. [CrossRef]
48. Jason, L.A.; Torres-Harding, S.R.; Carrico, A.W.; Taylor, R.R. Symptom occurrence in persons with chronic fatigue syndrome. *Biol. Psychol.* **2002**, *59*, 15–27. [CrossRef]
49. Stussman, B.; Williams, A.; Snow, J.; Gavin, A.; Scott, R.; Nath, A.; Walitt, B. Characterization of post-exertional malaise in patients with myalgic encephalomyelitis/chronic fatigue syndrome. *Front. Neurol.* **2020**, *11*, 1025. [CrossRef]
50. Van Ness, J.M.; Stevens, S.R.; Bateman, L.; Stiles, T.L.; Snell, C.R. Postexertional malaise in women with chronic fatigue syndrome. *J. Womens Health* **2010**, *19*, 239–244. [CrossRef]
51. O'Brien, K.K.; Brown, D.A.; Bergin, C.; Erlandson, K.M.; Vera, J.H.; Avery, L.; Carusone, S.C.; Cheung, A.M.; Goulding, S.; Harding, R.; et al. Long COVID and episodic disability: Advancing the conceptualisation, measurement and knowledge of episodic disability among people living with Long COVID—Protocol for a mixed-methods study. *BMJ Open* **2022**, *12*, e060826. [CrossRef] [PubMed]
52. O'Brien, K.K.; Brown, D.A.; McDuff, K.; St Clair-Sullivan, N.; Solomon, P.; Chan Carusone, S.; McCorkell, L.; Wei, H.; Goulding, S.; O'Hara, M.; et al. Conceptualising the episodic nature of disability among adults living with Long COVID: A qualitative study. *BMJ Glob. Health* **2023**, *8*, e011276. [CrossRef] [PubMed]
53. Sharpe, M.C.; Archard, L.C.; Banatvala, J.E.; Borysiewicz, L.K.; Clare, A.W.; David, A.; Edwards, R.H.; Hawton, K.E.; Lambert, H.P.; Lane, R.J.; et al. A report–chronic fatigue syndrome: Guidelines for research. *J. R. Soc. Med.* **1991**, *84*, 118–121. [CrossRef]

54. Reeves, W.C.; Wagner, D.; Nisenbaum, R.; Jones, J.F.; Gurbaxani, B.; Solomon, L.; Papanicolaou, D.A.; Unger, E.R.; Vernon, S.D.; Heim, C. Chronic fatigue syndrome—A clinically empirical approach to its definition and study. *BMC Med.* **2005**, *3*, 19. [CrossRef] [PubMed]
55. United States National Academy of Medicine. *Beyond Myalgic Encephalomyelitis/Chronic Fatigue Syndrome: Redefining an Illness. The National Academies Collection: Reports Funded by National Institutes of Health*; NAM: Washington, DC, USA, 2015.
56. National Institute for Health and Care Excellence. *Myalgic Encephalomyelitis (or Encephalopathy)/Chronic Fatigue Syndrome: Diagnosis and Management—NICE Guideline, No. 206*; National Institute for Health and Care Excellence: Guidelines; National Institute for Health and Care Excellence: London, UK, 2021.
57. A new clinical entity? *Lancet* **1956**, *270*, 789–790.
58. World Health Organization. *Manual of the International Statistical Classification of Diseases, Injuries, and Causes of Death Based on the Recommendations of the Eighth Revision Conference*, 8th ed.; World Health Organization: Geneva, Switzerland, 1969.
59. World Health Organization. *ICD-11: International Classification of Diseases, 11th revision*; World Health Organization: Geneva, Switzerland, 2022.
60. American Psychiatric Association. Committee on Nomenclature and Statistics. In *Mental Disorders*; Diagnostic and Statistical Manual; American Psychiatric Association: Washington, DC, USA, 1952.
61. American Psychiatric Association. Committee on Nomenclature and Statistics. In *Diagnostic and Statistical Manual of Mental Disorders*, 2nd ed.; American Psychiatric Association: Washington, DC, USA, 1968.
62. American Psychiatric Association. *Diagnostic and Statistical Manual of Mental Disorders*, 3rd ed.; American Psychiatric Association: Washington, DC, USA, 1980.
63. Owens, C.; Dein, S. Conversion disorder: The modern hysteria. *Adv. Psychiatr. Treat.* **2006**, *12*, 152–157. [CrossRef]
64. American Psychiatric Association. *Diagnostic and Statistical Manual of Mental Disorders*, 5th ed.; American Psychiatric Association: Arlington, VA, USA, 2013.
65. American Psychiatric Association. *Diagnostic and Statistical Manual of Mental Disorders: DSM-5-TR*, 5th ed., text revision ed.; American Psychiatric Association Publishing: Washington, DC, USA, 2022.
66. Mayo Clinic. Dissociative Disorders. Available online: https://www.mayoclinic.org/diseases-conditions/dissociative-disorders/symptoms-causes/syc-20355215 (accessed on 12 October 2023).
67. Diez, I.; Williams, B.; Kubicki, M.R.; Makris, N.; Perez, D.L. Reduced limbic microstructural integrity in functional neurological disorder. *Psychol. Med.* **2021**, *51*, 485–493. [CrossRef] [PubMed]
68. Hassa, T.; Spiteri, S.; Schmidt, R.; Merkel, C.; Schoenfeld, M.A. Increased amygdala activity associated with cognitive reappraisal strategy in functional neurologic disorder. *Front. Psychiatry* **2021**, *12*, 613156. [CrossRef] [PubMed]
69. Ospina, J.P.; Jalilianhasanpour, R.; Perez, D.L. The role of the anterior and midcingulate cortex in the neurobiology of functional neurologic disorder. *Handb. Clin. Neurol.* **2019**, *166*, 267–279. [CrossRef] [PubMed]
70. Perez, D.L.; Matin, N.; Williams, B.; Tanev, K.; Makris, N.; LaFrance, W.C., Jr.; Dickerson, B.C. Cortical thickness alterations linked to somatoform and psychological dissociation in functional neurological disorders. *Hum. Brain Mapp.* **2018**, *39*, 428–439. [CrossRef] [PubMed]
71. Perez, D.L.; Williams, B.; Matin, N.; LaFrance, W.C., Jr.; Costumero-Ramos, V.; Fricchione, G.L.; Sepulcre, J.; Keshavan, M.S.; Dickerson, B.C. Corticolimbic structural alterations linked to health status and trait anxiety in functional neurological disorder. *J. Neurol. Neurosurg. Psychiatry* **2017**, *88*, 1052–1059. [CrossRef] [PubMed]
72. Williams, B.; Jalilianhasanpour, R.; Matin, N.; Fricchione, G.L.; Sepulcre, J.; Keshavan, M.S.; LaFrance, W.C., Jr.; Dickerson, B.C.; Perez, D.L. Individual differences in corticolimbic structural profiles linked to insecure attachment and coping styles in motor functional neurological disorders. *J. Psychiatr. Res.* **2018**, *102*, 230–237. [CrossRef]
73. Gutkin, M.; McLean, L.; Brown, R.; Kanaan, R.A. Systematic review of psychotherapy for adults with functional neurological disorder. *J. Neurol. Neurosurg. Psychiatry* **2020**. *Online ahead of print*. [CrossRef] [PubMed]
74. Stone, J.; LaFrance, W.C., Jr.; Levenson, J.L.; Sharpe, M. Issues for DSM-5: Conversion disorder. *Am. J. Psychiatry* **2010**, *167*, 626–627. [CrossRef] [PubMed]
75. Stone, J.; Hoeritzauer, I.; McWhirter, L.; Carson, A. Functional neurological disorder: Defying dualism. *World Psychiatry* **2024**, *23*, 53–54. [CrossRef] [PubMed]
76. McLoughlin, C.; Hoeritzauer, I.; Cabreira, V.; Aybek, S.; Adams, C.; Alty, J.; Ball, H.A.; Baker, J.; Bullock, K.; Burness, C.; et al. Functional neurological disorder is a feminist issue. *J. Neurol. Neurosurg. Psychiatry* **2023**, *94*, 855–862. [CrossRef] [PubMed]
77. Stone, J.; Carson, A.; Sharpe, M. Functional symptoms and signs in neurology: Assessment and diagnosis. *J. Neurol. Neurosurg. Psychiatry* **2005**, *76* (Suppl. S1), i2–i12. [CrossRef] [PubMed]
78. National Institute for Health and Care Excellence. Suspected Neurological Conditions: Recognition and Referral | NICE Guideline [NG127]. Available online: https://www.nice.org.uk/guidance/ng127 (accessed on 12 October 2023).
79. ME Association. ME Association Petition: M.E. is Not a Functional Disorder. 27 September 2017. Available online: https://meassociation.org.uk/2017/09/me-association-petition-m-e-is-not-a-functional-disorder-27-september-2017/ (accessed on 12 October 2023).
80. Anonymous. The On-Call Doctor Rejected Line's Desperate Call Just Days before Her Husband's Death: 'He Was Let Down by the System'. Available online: https://www-bt-dk.translate.goog/samfund/vagtlaegen-afviste-lines-desperate-opkald-

faa-dage-foer-sin-mands-doed-han-blev?_x_tr_sl=da&_x_tr_tl=en&_x_tr_hl=en-US&_x_tr_pto=wapp (accessed on 12 October 2023).
81. ME Action. Advocacy Update from Finland. Available online: https://www.meaction.net/2020/12/07/advocacy-update-from-finland/ (accessed on 12 October 2023).
82. American Psychiatric Association. *Desk Reference to the Diagnostic Criteria from DSM-5-TR*; American Psychiatric Association: Washington, DC, USA, 2022.
83. Mishra, A.; Pandey, S. Functional neurological disorders: Clinical spectrum, diagnosis, and treatment. *Neurologist* **2022**, *27*, 276–289. [CrossRef] [PubMed]
84. Hira, R.; Karalasingham, K.; Baker, J.R.; Raj, S.R. Autonomic manifestations of long-COVID syndrome. *Curr. Neurol. Neurosci. Rep.* **2023**, *23*, 881–892. [CrossRef] [PubMed]
85. Turner, S.; Khan, M.A.; Putrino, D.; Woodcock, A.; Kell, D.B.; Pretorius, E. Long COVID: Pathophysiological factors and abnormalities of coagulation. *Trends Endocrinol. Metab.* **2023**, *34*, 321–344. [CrossRef]
86. Boccatonda, A.; Campello, E.; Simion, C.; Simioni, P. Long-term hypercoagulability, endotheliopathy and inflammation following acute SARS-CoV-2 infection. *Expert Rev. Hematol.* **2023**, *16*, 1035–1048. [CrossRef]
87. Smadja, D.M.; Mentzer, S.J.; Fontenay, M.; Laffan, M.A.; Ackermann, M.; Helms, J.; Jonigk, D.; Chocron, R.; Pier, G.B.; Gendron, N.; et al. COVID-19 is a systemic vascular hemopathy: Insight for mechanistic and clinical aspects. *Angiogenesis* **2021**, *24*, 755–788. [CrossRef] [PubMed]
88. Proal, A.D.; VanElzakker, M.B.; Aleman, S.; Bach, K.; Boribong, B.P.; Buggert, M.; Cherry, S.; Chertow, D.S.; Davies, H.E.; Dupont, C.L.; et al. SARS-CoV-2 reservoir in post-acute sequelae of COVID-19 (PASC). *Nat. Immunol.* **2023**, *24*, 1616–1627. [CrossRef] [PubMed]
89. Yin, K.; Peluso, M.J.; Luo, X.; Thomas, R.; Shin, M.G.; Neidleman, J.; Andrew, A.; Young, K.C.; Ma, T.; Hoh, R.; et al. Long COVID manifests with T cell dysregulation, inflammation and an uncoordinated adaptive immune response to SARS-CoV-2. *Nat. Immunol.* **2024**, *25*, 218–225. [CrossRef]
90. Roe, K. A role for T-cell exhaustion in Long COVID-19 and severe outcomes for several categories of COVID-19 patients. *J. Neurosci. Res.* **2021**, *99*, 2367–2376. [CrossRef] [PubMed]
91. Durstenfeld, M.S.; Peluso, M.J.; Kaveti, P.; Hill, C.; Li, D.; Sander, E.; Swaminathan, S.; Arechiga, V.M.; Lu, S.; Goldberg, S.A.; et al. Reduced exercise capacity, chronotropic incompetence, and early systemic inflammation in cardiopulmonary phenotype long coronavirus disease 2019. *J. Infect. Dis.* **2023**, *228*, 542–554. [CrossRef] [PubMed]
92. Durstenfeld, M.S.; Sun, K.; Tahir, P.; Peluso, M.J.; Deeks, S.G.; Aras, M.A.; Grandis, D.J.; Long, C.S.; Beatty, A.; Hsue, P.Y. Use of cardiopulmonary exercise testing to evaluate long COVID-19 symptoms in adults: A systematic review and meta-analysis. *JAMA Netw. Open* **2022**, *5*, e2236057. [CrossRef] [PubMed]
93. Ortona, E.; Buonsenso, D.; Carfi, A.; Malorni, W.; Long COVID Kids study group. Long COVID: An estrogen-associated autoimmune disease? *Cell Death Discov.* **2021**, *7*, 77. [CrossRef] [PubMed]
94. Amiral, J.; Seghatchian, J. Autoimmune complications of COVID-19 and potential consequences for long-lasting disease syndromes. *Transfus. Apher. Sci.* **2023**, *62*, 103625. [CrossRef] [PubMed]
95. Vojdani, A.; Vojdani, E.; Saidara, E.; Maes, M. Persistent SARS-CoV-2 Infection, EBV, HHV-6 and other factors may contribute to inflammation and autoimmunity in long COVID. *Viruses* **2023**, *15*, 400. [CrossRef] [PubMed]
96. Kahn, P.A.; Joseph, P.; Heerdt, P.M.; Singh, I. Differential cardiopulmonary haemodynamic phenotypes in PASC-related exercise intolerance. *ERJ Open Res.* **2024**, *10*, 00714-2023. [CrossRef] [PubMed]
97. Singh, I.; Joseph, P.; Heerdt, P.M.; Cullinan, M.; Lutchmansingh, D.D.; Gulati, M.; Possick, J.D.; Systrom, D.M.; Waxman, A.B. Persistent exertional intolerance after COVID-19: Insights from invasive cardiopulmonary exercise testing. *Chest* **2022**, *161*, 54–63. [CrossRef] [PubMed]
98. Appelman, B.; Charlton, B.T.; Goulding, R.P.; Kerkhoff, T.J.; Breedveld, E.A.; Noort, W.; Offringa, C.; Bloemers, F.W.; van Weeghel, M.; Schomakers, B.V.; et al. Muscle abnormalities worsen after post-exertional malaise in long COVID. *Nat. Commun.* **2024**, *15*, 17. [CrossRef] [PubMed]
99. McAlpine, L.; Zubair, A.S.; Joseph, P.; Spudich, S. Case-control study of individuals with small fiber neuropathy after COVID-19. *Neurol. Neuroimmunol. Neuroinflamm.* **2024**, *11*, e200244. [CrossRef] [PubMed]
100. Nagata, N.; Takeuchi, T.; Masuoka, H.; Aoki, R.; Ishikane, M.; Iwamoto, N.; Sugiyama, M.; Suda, W.; Nakanishi, Y.; Terada-Hirashima, J.; et al. Human gut microbiota and its metabolites impact immune responses in COVID-19 and its complications. *Gastroenterology* **2023**, *164*, 272–288. [CrossRef] [PubMed]
101. United States Centers for Disease Control and Prevention. New ICD-10-CM Code for Post-COVID Conditions, Following the 2019 Novel Coronavirus (COVID-19). Available online: https://www.cdc.gov/nchs/data/icd/announcement-new-icd-code-for-post-covid-condition-april-2022-final.pdf (accessed on 10 July 2023).
102. Novakova, L.; Anyz, J.; Forejtova, Z.; Rosikova, T.; Vechetova, G.; Sojka, P.; Ruzicka, E.; Serranova, T. Increased frequency of self-reported obsessive-compulsive symptoms in patients with functional movement disorders. *Mov. Disord. Clin. Pract.* **2023**, *10*, 1341–1348. [CrossRef]
103. Roivainen, E.; Peura, M.; Patsi, J. Cognitive profile in functional disorders. *Cogn. Neuropsychiatry* **2023**, *28*, 424–436. [CrossRef]
104. Hamilton, J.; Campos, R.; Creed, F. Anxiety, depression and management of medically unexplained symptoms in medical clinics. *J. R. Coll. Physicians Lond.* **1996**, *30*, 18–20. [PubMed]

105. Afari, N.; Ahumada, S.M.; Wright, L.J.; Mostoufi, S.; Golnari, G.; Reis, V.; Cuneo, J.G. Psychological trauma and functional somatic syndromes: A systematic review and meta-analysis. *Psychosom. Med.* **2014**, *76*, 2–11. [CrossRef] [PubMed]
106. Teodoro, T.; Chen, J.; Gelauff, J.; Edwards, M.J. Functional neurological disorder in people with long COVID: A systematic review. *Eur. J. Neurol.* **2023**, *30*, 1505–1514. [CrossRef]
107. Bennett, K.; Diamond, C.; Hoeritzauer, I.; Gardiner, P.; McWhirter, L.; Carson, A.; Stone, J. A practical review of functional neurological disorder (FND) for the general physician. *Clin. Med.* **2021**, *21*, 28–36. [CrossRef] [PubMed]
108. Oaklander, A.L.; Mills, A.J.; Kelley, M.; Toran, L.S.; Smith, B.; Dalakas, M.C.; Nath, A. Peripheral neuropathy evaluations of patients with prolonged long COVID. *Neurol. Neuroimmunol. Neuroinflamm.* **2022**, *9*, e1146. [CrossRef] [PubMed]
109. Gemignani, F.; Bellanova, M.F.; Saccani, E.; Pavesi, G. Non-length-dependent small fiber neuropathy: Not a matter of stockings and gloves. *Muscle Nerve* **2022**, *65*, 10–28. [CrossRef] [PubMed]
110. Larsen, N.W.; Stiles, L.E.; Shaik, R.; Schneider, L.; Muppidi, S.; Tsui, C.T.; Geng, L.N.; Bonilla, H.; Miglis, M.G. Characterization of autonomic symptom burden in long COVID: A global survey of 2314 adults. *Front. Neurol.* **2022**, *13*, 1012668. [CrossRef] [PubMed]
111. Thijs, R.D.; Brignole, M.; Falup-Pecurariu, C.; Fanciulli, A.; Freeman, R.; Guaraldi, P.; Jordan, J.; Habek, M.; Hilz, M.; Traon, A.P.; et al. Recommendations for tilt table testing and other provocative cardiovascular autonomic tests in conditions that may cause transient loss of consciousness: Consensus statement of the European Federation of Autonomic Societies (EFAS) endorsed by the American Autonomic Society (AAS) and the European Academy of Neurology (EAN). *Clin. Auton. Res.* **2021**, *31*, 369–384. [CrossRef] [PubMed]
112. Panicker, J.N.; Selai, C.; Herve, F.; Rademakers, K.; Dmochowski, R.; Tarcan, T.; von Gontard, A.; Vrijens, D. Psychological comorbidities and functional neurological disorders in women with idiopathic urinary retention: International Consultation on Incontinence Research Society (ICI-RS) 2019. *Neurourol. Urodyn.* **2020**, *39* (Suppl. S3), S60–S69. [CrossRef] [PubMed]
113. Shouman, K.; Vanichkachorn, G.; Cheshire, W.P.; Suarez, M.D.; Shelly, S.; Lamotte, G.J.; Sandroni, P.; Benarroch, E.E.; Berini, S.E.; Cutsforth-Gregory, J.K.; et al. Autonomic dysfunction following COVID-19 infection: An early experience. *Clin. Auton. Res.* **2021**, *31*, 385–394. [CrossRef] [PubMed]
114. McWhirter, L.; Ritchie, C.; Stone, J.; Carson, A. Functional cognitive disorders: A systematic review. *Lancet Psychiatry* **2020**, *7*, 191–207. [CrossRef] [PubMed]
115. Monje, M.; Iwasaki, A. The neurobiology of long COVID. *Neuron* **2022**, *110*, 3484–3496. [CrossRef] [PubMed]
116. Soung, A.L.; Vanderheiden, A.; Nordvig, A.S.; Sissoko, C.A.; Canoll, P.; Mariani, M.B.; Jiang, X.; Bricker, T.; Rosoklija, G.B.; Arango, V.; et al. COVID-19 induces CNS cytokine expression and loss of hippocampal neurogenesis. *Brain* **2022**, *145*, 4193–4201. [CrossRef]
117. Voruz, P.; Allali, G.; Benzakour, L.; Nuber-Champier, A.; Thomasson, M.; Jacot de Alcântara, I.; Pierce, J.; Lalive, P.H.; Lövblad, K.-O.; Braillard, O.; et al. Long COVID neuropsychological deficits after severe, moderate, or mild infection. *Clin. Transl. Neurosci.* **2022**, *6*, 9. [CrossRef]
118. Blackmon, K.; Day, G.S.; Powers, H.R.; Bosch, W.; Prabhakaran, D.; Woolston, D.; Pedraza, O. Neurocognitive screening in patients following SARS-CoV-2 infection: Tools for triage. *BMC Neurol.* **2022**, *22*, 285. [CrossRef] [PubMed]
119. Lynch, S.; Ferrando, S.J.; Dornbush, R.; Shahar, S.; Smiley, A.; Klepacz, L. Screening for brain fog: Is the montreal cognitive assessment an effective screening tool for neurocognitive complaints post-COVID-19? *Gen. Hosp. Psychiatry* **2022**, *78*, 80–86. [CrossRef] [PubMed]
120. Klein, J.; Wood, J.; Jaycox, J.R.; Dhodapkar, R.M.; Lu, P.; Gehlhausen, J.R.; Tabachnikova, A.; Greene, K.; Tabacof, L.; Malik, A.A.; et al. Distinguishing features of long COVID identified through immune profiling. *Nature* **2023**, *623*, 139–148. [CrossRef] [PubMed]
121. Altmann, D.M.; Whettlock, E.M.; Liu, S.; Arachchillage, D.J.; Boyton, R.J. The immunology of long COVID. *Nat. Rev. Immunol.* **2023**, *23*, 618–634. [CrossRef] [PubMed]
122. Ruffieux, H.; Hanson, A.L.; Lodge, S.; Lawler, N.G.; Whiley, L.; Gray, N.; Nolan, T.H.; Bergamaschi, L.; Mescia, F.; Turner, L.; et al. A patient-centric modeling framework captures recovery from SARS-CoV-2 infection. *Nat. Immunol.* **2023**, *24*, 349–358. [CrossRef] [PubMed]
123. Mina, Y.; Enose-Akahata, Y.; Hammoud, D.A.; Videckis, A.J.; Narpala, S.R.; O'Connell, S.E.; Carroll, R.; Lin, B.C.; McMahan, C.C.; Nair, G.; et al. Deep phenotyping of neurologic postacute sequelae of SARS-CoV-2 infection. *Neurol. Neuroimmunol. Neuroinflamm.* **2023**, *10*, e200097. [CrossRef] [PubMed]
124. Douaud, G.; Lee, S.; Alfaro-Almagro, F.; Arthofer, C.; Wang, C.; McCarthy, P.; Lange, F.; Andersson, J.L.R.; Griffanti, L.; Duff, E.; et al. SARS-CoV-2 is associated with changes in brain structure in UK Biobank. *Nature* **2022**, *604*, 697–707. [CrossRef] [PubMed]
125. Hosp, J.A.; Reisert, M.; Dressing, A.; Gotz, V.; Kellner, E.; Mast, H.; Arndt, S.; Waller, C.F.; Wagner, D.; Rieg, S.; et al. Cerebral microstructural alterations in Post-COVID-condition are related to cognitive impairment, olfactory dysfunction and fatigue. *Nat. Commun.* **2024**, *15*, 4256. [CrossRef] [PubMed]
126. Wu, L.; Zhang, Z.; Liang, X.; Wang, Y.; Cao, Y.; Li, M.; Zhou, F. Glymphatic system dysfunction in recovered patients with mild COVID-19: A DTI-ALPS study. *iScience* **2024**, *27*, 108647. [CrossRef] [PubMed]
127. Greene, C.; Connolly, R.; Brennan, D.; Laffan, A.; O'Keeffe, E.; Zaporojan, L.; O'Callaghan, J.; Thomson, B.; Connolly, E.; Argue, R.; et al. Blood-brain barrier disruption and sustained systemic inflammation in individuals with long COVID-associated cognitive impairment. *Nat. Neurosci.* **2024**, *27*, 421–432. [CrossRef] [PubMed]

128. Chaganti, J.; Poudel, G.; Cysique, L.A.; Dore, G.J.; Kelleher, A.; Matthews, G.; Darley, D.; Byrne, A.; Jakabek, D.; Zhang, X.; et al. Blood brain barrier disruption and glutamatergic excitotoxicity in post-acute sequelae of SARS COV-2 infection cognitive impairment: Potential biomarkers and a window into pathogenesis. *Front. Neurol.* **2024**, *15*, 1350848. [CrossRef] [PubMed]
129. VanElzakker, M.B.; Bues, H.F.; Brusaferri, L.; Kim, M.; Saadi, D.; Ratai, E.M.; Dougherty, D.D.; Loggia, M.L. Neuroinflammation in post-acute sequelae of COVID-19 (PASC) as assessed by [(11)C]PBR28 PET correlates with vascular disease measures. *Brain Behav. Immun.* **2024**, *119*, 713–723. [CrossRef]
130. Peluso, M.J.; Ryder, D.; Flavell, R.R.; Wang, Y.; Levi, J.; LaFranchi, B.H.; Deveau, T.M.; Buck, A.M.; Munter, S.E.; Asare, K.A.; et al. Tissue-based T cell activation and viral RNA persist for up to 2 years after SARS-CoV-2 infection. *Sci. Transl. Med.* **2024**, *16*, eadk3295. [CrossRef] [PubMed]
131. Perez, D.L.; Nicholson, T.R.; Asadi-Pooya, A.A.; Begue, I.; Butler, M.; Carson, A.J.; David, A.S.; Deeley, Q.; Diez, I.; Edwards, M.J.; et al. Neuroimaging in functional neurological disorder: State of the field and research agenda. *Neuroimage Clin.* **2021**, *30*, 102623. [CrossRef] [PubMed]
132. Mavroudis, I.; Kazis, D.; Kamal, F.Z.; Gurzu, I.L.; Ciobica, A.; Padurariu, M.; Novac, B.; Iordache, A. Understanding Functional Neurological Disorder: Recent insights and diagnostic challenges. *Int. J. Mol. Sci.* **2024**, *25*, 4470. [CrossRef] [PubMed]
133. Schneider, A.; Weber, S.; Wyss, A.; Loukas, S.; Aybek, S. BOLD signal variability as potential new biomarker of functional neurological disorders. *Neuroimage Clin.* **2024**, *43*, 103625. [CrossRef] [PubMed]
134. Maurer, C.W.; LaFaver, K.; Limachia, G.S.; Capitan, G.; Ameli, R.; Sinclair, S.; Epstein, S.A.; Hallett, M.; Horovitz, S.G. Gray matter differences in patients with functional movement disorders. *Neurology* **2018**, *91*, e1870–e1879. [CrossRef] [PubMed]
135. Begue, I.; Adams, C.; Stone, J.; Perez, D.L. Structural alterations in functional neurological disorder and related conditions: A software and hardware problem? *Neuroimage Clin.* **2019**, *22*, 101798. [CrossRef] [PubMed]
136. van Campen, C.; Rowe, P.C.; Verheugt, F.W.A.; Visser, F.C. Cognitive function declines following orthostatic stress in adults with myalgic encephalomyelitis/chronic fatigue syndrome (ME/CFS). *Front. Neurosci.* **2020**, *14*, 688. [CrossRef] [PubMed]
137. van Campen, C.; Rowe, P.C.; Verheugt, F.W.A.; Visser, F.C. Numeric rating scales show prolonged post-exertional symptoms after orthostatic testing of adults with myalgic encephalomyelitis/chronic fatigue syndrome. *Front. Med.* **2020**, *7*, 602894. [CrossRef] [PubMed]
138. van Campen, C.; Rowe, P.C.; Visser, F.C. Cerebral blood flow is reduced in severe myalgic encephalomyelitis/chronic fatigue syndrome patients during mild orthostatic stress testing: An exploratory study at 20 degrees of head-up tilt testing. *Healthcare* **2020**, *8*, 169. [CrossRef] [PubMed]
139. van Campen, C.; Rowe, P.C.; Visser, F.C. Cerebral blood flow remains reduced after tilt testing in myalgic encephalomyelitis/chronic fatigue syndrome patients. *Clin. Neurophysiol. Pract.* **2021**, *6*, 245–255. [CrossRef] [PubMed]
140. van Campen, C.M.C.; Rowe, P.C.; Visser, F.C. Worsening symptoms is associated with larger cerebral blood flow abnormalities during tilt-testing in myalgic encephalomyelitis/chronic fatigue syndrome (ME/CFS). *Medicina* **2023**, *59*, 2153. [CrossRef] [PubMed]
141. Blitshteyn, S.; Verduzco-Gutierrez, M. Long COVID: A major public health issue. *Am. J. Phys. Med. Rehabil.* 2024, *Online ahead of print*. [CrossRef] [PubMed]

Journal of
Personalized Medicine

MDPI

Article

Dysautonomia, but Not Cardiac Dysfunction, Is Common in a Cohort of Individuals with Long COVID

Laura Tabacof [1,†], Jamie Wood [1,†], Erica Breyman [1], Jenna Tosto-Mancuso [1], Amanda Kelly [2], Kaitlyn Wilkey [2], Chi Zhang [3], David Putrino [1,*,†] and Amy Kontorovich [3,4,5,†]

[1] Abilities Research Center, Department of Rehabilitation and Human Performance, Icahn School of Medicine at Mount Sinai, New York, NY 10029, USA; jamie.wood@mountsinai.org (J.W.)
[2] Department of Rehabilitation and Human Performance, Icahn School of Medicine at Mount Sinai, New York, NY 10029, USA
[3] Zena and Michael A. Weiner Cardiovascular Institute, Icahn School of Medicine at Mount Sinai, New York, NY 10029, USA; chi.zhang2@mountsinai.org (C.Z.); amy.kontorovich@mountsinai.org (A.K.)
[4] The Mindich Child Health and Development Institute, Icahn School of Medicine at Mount Sinai, New York, NY 10029, USA
[5] The Cardiovascular Research Institute, Icahn School of Medicine at Mount Sinai, New York, NY 10029, USA
[*] Correspondence: david.putrino@mountsinai.org
[†] These authors contributed equally to this work.

Abstract: Despite the prevalence of dysautonomia in people with Long COVID, it is currently unknown whether Long COVID dysautonomia is routinely accompanied by structural or functional cardiac alterations. In this retrospective observational study, the presence of echocardiographic abnormalities was assessed. Left ventricular (LV) chamber sizes were correlated to diagnostic categories and symptoms via standardized patient-reported outcome (PRO) questionnaires. A total of 203 individuals with Long COVID without pre-existing cardiac disease and with available echocardiograms were included (mean age, 45 years; 67% female). Overall, symptoms and PRO scores for fatigue, breathlessness, quality of life, disability, anxiety and depression were not different between those classified with post-COVID dysautonomia (PCD, 22%) and those unclassified (78%). An LV internal diameter at an end-diastole z score < -2 was observed in 33 (16.5%) individuals, and stroke volume (SV) was lower in the PCD vs. unclassified subgroup (51.6 vs. 59.2 mL, 95% C.I. 47.1–56.1 vs. 56.2–62.3). LV end-diastolic volume (mean diff. (95% CI) -13 [-1–-26] mL, $p = 0.04$) and SV (-10 [-1–-20] mL, $p = 0.03$) were smaller in those individuals reporting a reduction in physical activity post-COVID-19 infection, and smaller LVMI was weakly correlated with worse fatigue ($r = 0.23$, $p = 0.02$). The majority of individuals with Long COVID report shared symptoms and did not demonstrate cardiac dysfunction on echocardiography.

Keywords: postural orthostatic tachycardia syndrome (POTS); dysautonomia; long COVID; echocardiogram; post-acute sequelae of COVID; PASC

1. Introduction

Although SARS-CoV-2 infection has been phenomenologically linked with both cardiac abnormalities [1–8] and persistent lingering symptoms referred to as Long COVID [9,10], it remains unknown whether these two sets of complications are pathologically connected in a majority of cases. Cardiac effects from COVID-19 infection have been described within several categories: (1) myocardial injury seen in ~30% of acutely hospitalized individuals, as evidenced by elevated cardiac biomarkers (i.e., troponin) [1]; (2) coagulopathy leading to microvascular or macrovascular thromboembolic events [2,3]; (3) rare cases of clinical myocarditis or myopericarditis necessitating hospitalization either in adults or in children in the context of Multisystem Inflammatory Syndrome in Children (MIS-C) [4–6]; and (4) cardiac MRI evidence of inflammation, identified primarily in prospective surveys of those with mild initial COVID-19 illness, such as young athletes, which is typically limited and transient [7,8]. In distinct contrast to these COVID-related cardiac endotypes, Long COVID is estimated to affect as many as 20% of individuals who have survived an initial acute COVID-19 infection [11] and leads to diverse chronic symptoms, typically without objective abnormalities on standard-of-care diagnostic laboratory or imaging tests [9]. Symptoms include those that may be considered referable to the cardiovascular system (chest discomfort, palpitations, breathlessness and exercise intolerance) as well as systemic (fatigue, brain fog, dizziness, memory loss and weakness) [9,10]. Many individuals experience debilitating symptoms for more than 12 months following acute infection [9].

A cornerstone of Long COVID is post-exertional symptom exacerbation following an increase in levels of physical or mental exertion [9,10]. This feature is also cardinal to a set of conditions that fall within the classification of autonomic nervous system dysfunction or dysautonomia. Indeed, dysautonomia has been described in Long COVID [12–15], and the reported causes of symptom exacerbation (physical exertion, stress, dehydration, weather changes, consuming large meals, premenstrual period and alcohol) and reductions in levels of physical activity are shared between Long COVID and other forms of dysautonomia [9,10,12,16,17]. Most individual presentations of dysautonomia in Long COVID span multiple subtypes of autonomic dysfunction, with up to 33% meeting criteria for specific diagnoses such as postural orthostatic tachycardia syndrome (POTS) [18–20]. Dysautonomia is not classically associated with myocardial inflammation or dysfunction. However, given the prevalence of myocardial involvement as a sequela of COVID-19, it was sought to determine whether post-acute COVID dysautonomia (PCD) is associated with distinct cardiac abnormalities, including features of cardiac atrophy, as reported in other forms of dysautonomia [16,17]. In this study, symptom burden and echocardiographic findings among a "real world," richly phenotyped Long COVID/post-acute COVID-19 dysautonomia (PCD) cohort were analyzed retrospectively.

This report describes the findings from routine echocardiogram assessments obtained from a cohort of individuals attending a Long COVID clinic, and explores the presence of clinically diagnosed PCD alongside self-reported persistent symptoms.

2. Materials and Methods

2.1. Study Design

This was an observational study using retrospectively obtained electronic health record (EHR) information and patient-reported outcomes (PROs). Approval for publication was provided by the Mount Sinai Program for Protection of Human Subjects (IRB 21-00944). A waiver of consent was approved.

2.2. Participants

Adults attending the Long COVID clinic at Mount Sinai Hospital were included. The Long COVID clinic is an interdisciplinary clinic consisting of physicians (primary care and a range of subspecialties including physiatry and cardiology), physical therapists, dietitians and researchers.

Inclusion criteria were EHR diagnosis of Long COVID, defined as experiencing new, returning or ongoing health problems 4 or more weeks following initial COVID-19 infection in the absence of any specific organ damage using standard clinical testing protocols [9,21], and having a transthoracic echocardiogram (TTE) assessment performed at Mount Sinai Hospital > 28 days following diagnosis with COVID-19. Individuals were excluded if they had a diagnosis of heart failure, cardiomyopathy or dysautonomia prior to COVID-19 infection.

2.3. Data Collection and Outcomes

Data including demographics, acute COVID-19 hospitalization status, need for mechanical ventilation and duration of COVID-19-related symptoms were obtained retrospectively from the EHRs as well as a patient-reported outcome (PRO) surveys developed by Long COVID clinic team members and administered as part of clinical care.

2.4. Echocardiographic Assessment

Echocardiographic data were systematically extracted from clinical reports through a Mount Sinai Data Warehouse query and included left ventricular (LV) and right ventricular (RV) size and function; LV ejection fraction (EF); left and right atrial sizes; qualitative descriptors of the aortic, mitral, tricuspid and pulmonic valve anatomy and function; presence/absence of pulmonary hypertension; thoracic aortic dilatation; pericardial abnormalities; LV internal diameter at end diastole (LVIDd) and end systole (LVIDs); LV end-diastolic volume (LVEDV), end-systolic volume (LVESV) and stroke volume (SV) from the four-chamber apical view; and LV mass index (LVMI). The LVIDd and LVIDs z scores were calculated using the formula: $(LVID_{measured} - LVID_{mean})/SD$, with normal mean and SD values obtained from the American Society of Echocardiography and the European Association of Cardiovascular Imaging [22]. Additionally, the following quantification methods were used for other outcomes: Teicholz 2D (LVIDd) and biplane Simpson (LVEDV 4C, LVESV 4C, EF); SV 4C was calculated using 0.785 × left ventricular outflow tract (LVOT) diameter2 × LVOT velocity time integral; LV mass index was calculated using 0.8 (1.04 ([LVIDd + posterior wall thickness in diastole + interventricular septum thickness in diastole]3 − [LVIDd]3))+ 0.6 g; and LV mass index was calculated using LV mass/BSA.

Abnormalities of valves were defined as either structural abnormalities or moderate or higher valvular stenosis or insufficiency. Echocardiographic reports were manually over-read by the investigators (C.Z., A.K.) to ensure completeness of the data. Echocardiograms performed prior to COVID-19 infection were also reviewed if available.

2.5. Classification of Clinically Diagnosed Dysautonomia

EHRs (primary care or cardiology progress notes, "problem list") were manually reviewed to identify whether individuals with Long COVID were diagnosed with dysautonomia/PCD specifically as documented by their treating cardiologist or physician or were otherwise "unclassified" (no stated diagnosis of dysautonomia in the EHR). Participants were classified by the research team as having dysautonomia if their treating cardiologist or physician documented this diagnosis in the EHR, with supporting evidence including symptoms and clinical/historical features, and/or formal testing including a tilt table test, active stand test, quantitative sudomotor axon reflex test, thermoregulatory sweat test, or using heart rate variability. Members of the research team were not involved in the initial diagnosis of dysautonomia.

2.6. Patient-Reported Outcomes

Individuals attending the Long COVID clinic were requested to complete a PRO survey as part of their routine clinical care. Survey data were collected using Research Electronic Data Capture (REDCap) electronic data capture tools hosted in the Mount Sinai Health System. Participants were provided with a survey link via email to complete online. The PROs included persistent symptoms and triggers of symptom exacerbation and screening tools for breathlessness (Medical Research Council (MRC) Breathlessness Scale), health-related quality of life (HRQoL) (EuroQol EQ-5D-5L), fatigue (Fatigue Severity Scale (FSS), fatigue visual analog scale (VAS)), completion of regular-, moderate- and vigorous-intensity physical activity (author developed), cognitive function (Neuro-QOL), anxiety (GAD-7), depression (PHQ-2) and disability (WHODAS).

2.7. Statistical Plan

Statistical analyses were undertaken with Stata (StataCorp, Stata Statistical Software Release: V.14). Data are presented as frequencies and proportions, mean and standard deviation (SD) or median and 95% confidence interval (CI) where appropriate. Correlations between echocardiographic variables and PROs were examined using Pearson's correlation or Spearman's correlation where appropriate. Independent sample t-tests were used to examine between-group differences based on the classification of dysautonomia, presented as mean difference and 95% CI. Proportions were examined using Pearson's chi-squared test, presented as frequency (%).

3. Results

Participants were identified from a database of 737 individuals with Long COVID. Of these, 217 (29%) had an echocardiogram performed in a clinical setting at least 28 days following their COVID-19 infection (Figure 1). Fourteen (6%) were excluded from analyses as they had pre-existing (i.e., prior to COVID-19 infection) diagnosed heart failure, cardiomyopathy and/or dysautonomia. The majority of the final cohort were female (67%) and did not require hospitalization during acute COVID-19 infection (Table 1).

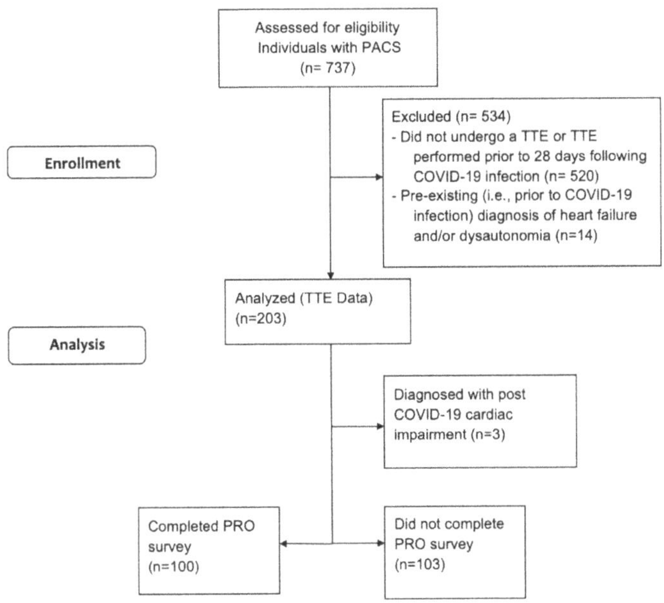

Figure 1. Consort flow diagram.

Table 1. Demographics of individuals with Long COVID (n = 203).

	All Participants (n = 203)	Female (n = 135)	Male (n = 68)	Dysautonomia (n = 45)	Unclassified (n = 158)
Female, n (%)	135 (67%)	135 (100)	0 (0)	37 (82)	98 (62)
Age, y	45 (22–80)	46 (22–79)	44 (23–80)	47 (23–79)	45 (22–80)
BSA	1.9 (1.3–2.6)	1.8 (1.3–2.5)	2.1 (1.6–2.6)	1.8 (1.4–2.3)	1.9 (1.3–2.6)
Hospitalized for COVID-19, n (%)	44 (22)	29 (21)	15 (22)	8 (18)	36 (23)
Mechanical ventilation, n (%)	5 (2)	5 (4)	0 (0)	0 (0)	8 (18)
Duration of symptoms, days	218 (34–500)	221 (36–479)	213 (34–500)	212 (36–479)	220 (34–500)

Data are presented as mean (range) otherwise indicated. BSA—body surface area. PCR—polymerase chain reaction. Mechanical ventilation—if required, mechanical ventilation was utilized during acute COVID-19 infection.

Echocardiographic abnormalities among the 203 included individuals are presented in Table 2. LV and RV systolic function was normal in 201 (99%) and 203 (100%) individuals, respectively. Of the 202 (99%) individuals for whom an EF was reported, the median (95% confidence interval (CI)) value was 65 ± 5%. Diastolic dysfunction was present in

8 (4%), LV hypertrophy in 20 patients (11%), pulmonary hypertension in 2 (1%; 1 also with RV dilatation) and pericardial abnormalities in 3 (1%) individuals. These abnormalities were new (n = 2, 7, 1, 2), documented prior to COVID-19 infection (n = 0, 2, 0, 1) or prior echocardiograms were unavailable (n = 6, 11, 1, 0) in affected individuals. Three (1%) individuals were diagnosed with myocardial impairments attributed to COVID-19 infection: one patient with segmental wall motion abnormality with borderline LV EF (no prior echocardiogram available for comparison), one patient with normal LV size and diffuse borderline LV systolic dysfunction (EF 50%; no prior echocardiogram available) and one with severe LV dilatation and EF 56% (new compared to pre-COVID-19 echocardiogram). Since Long COVID was defined as cases without structural cardiac changes that could otherwise explain persistent symptoms, these individuals were excluded from downstream analyses.

Table 2. Echocardiographic abnormalities in individuals with Long COVID (n = 203).

	Parameter Reported (%)	Abnormal (%)
LV systolic function	203 (100)	2 (1)
LV diastolic function	189 (93)	8 (4)
LV hypertrophy	179 (88)	20 (11)
LV dilatation	203 (100)	1 (0.5)
RV systolic function	197 (97)	0 (0)
RV size	197 (97)	2 (1)
Pulmonary hypertension	77 (38)	2 (3)
LA size	127 (63)	4 (3)
RA size	127 (63)	1 (1)
Mitral valve	200 (99)	1 (1)
Aortic valve	200 (99)	1 (1)
Tricuspid valve	200 (99)	0 (0)
Pulmonic valve	200 (99)	0 (0)
Pericardium	113 (56)	3 (3)
Aortic root dilatation	85 (42)	3 (4)
Miscellaneous	-	3 (1)

Data are reported as number and percentage. LV—left ventricle; RV—right ventricle; LA—left atrium; RA—right atrium. Mitral valve abnormality was mitral valve prolapse with mild-to-moderate insufficiency. Aortic valve abnormality was bicuspid aortic valve. Miscellaneous abnormalities included interatrial septal aneurysm (n = 1); lipomatous atrial septal hypertrophy (n = 1); ventricular septal defect versus sinus of Valsalva aneurysm (n = 1).

Of the 200 individuals with Long COVID included in the secondary analyses, 44 (22%) were classified by their provider (primary care or cardiologist) as having PCD; classifications were made based on symptoms and clinical/historical features (n = 28 (64%)) and/or formal testing (n = 17 (39%); these included tilt table test, active stand test, heart rate variability, quantitative sudomotor axon reflex test and/or thermoregulatory sweat test). The remaining 156 (78%) were labeled as having Long COVID without further classification.

Stroke volumes (SVs) were lower in the PCD group when compared to those not classified (Table 3). A similar subset of individuals in both the PCD (n = 10, 23%) and unclassified groups (n = 23, 15%) had LVIDd measurements smaller than sex-specific normal expected values (Z-score ≤ -2, $p = 0.25$) [22]. Overall, there were no differences in echocardiographic LV chamber size measures between the two groups.

Table 3. Echocardiographic measures in individuals with Long COVID (n = 200).

	All Participants (n = 200)	Dysautonomia (n = 44)	Unclassified (n = 156)	Difference
LVIDd z score	−0.89 (−1.06–−0.72)	−1.16 (−1.55–−0.77)	−0.81 (−1.00–−0.62)	0.35 (−0.06–0.72)
LVIDs z score [a]	−0.42 (−0.57–−0.28)	−0.50 (−0.82–−0.18)	−0.40 (−0.57–−0.18)	0.10 (−0.26–0.45)
LVEDV [a]	89.0 (85.3–92.5)	82.5 (75.5–89.4)	90.8 (86.6–94.9)	8.3 (−0.3–16.9)
LVESV [a]	39.3 (33.8–44.7)	30.0 (25.0–35.0)	42.2 (35.2–49.1)	12.2 (−0.5–24.9)
SV [a]	57.5 (54.9–60.1)	51.6 (47.1–56.1)	59.2 (56.2–62.3)	7.6 (1.6–13.7)
LVMI	85.2 (82.4–88.0)	85.8 (79.3–92.3)	85.0 (82.0–88.1)	0.8 (−7.5–5.9)

Data are presented as mean (95% confidence interval). [a] Echocardiographic measurements not available for all participants. LVID z score: dysautonomia, n = 43; not classified, n = 150. LVEDV: dysautonomia, n = 37; not classified, n = 130. LVESV: dysautonomia, n = 23; not classified, n = 73. SV: dysautonomia, n = 35; not classified, n = 116.

Of the 200 individuals with echocardiographic measures reported, 99 (50%) completed the PRO survey. The mean (95% CI) total number of symptoms reported was 12 (1 to 33) for individuals with PCD, as well as individuals who were unclassified. Scores from PROs screening for fatigue, breathlessness, quality of life, disability, anxiety and depression were not different between those with PCD and those who were unclassified (Table 4).

Table 4. Patient-reported outcomes for individuals (n = 99) who completed the survey.

Patient-Reported Outcome	All Participants (n = 99)	Dysautonomia (n = 27)	Unclassified (n = 72)	Difference (95% CI)
MRC breathlessness scale	2 (1–4)	2 (1–4)	2 (1–4)	0 (−1–0)
EQ-5D-5L domains				
Mobility	2 (1–4)	2 (1–4)	2 (1–4)	0 (−1–0)
Usual activities	3 (1–5)	3 (1–5)	3 (1–5)	0 (−1–1)
Anxiety/depression	3 (1–4)	3 (1–4)	3 (1–4)	0 (−1–1)
Self-care	1 (1–3)	1 (1–3)	1 (1–3)	0 (−1–0)
Pain/discomfort	3 (1–4)	3 (1–4)	2 (1–4)	−1 (−1–0)
EQ-5D-5L health status VAS [a,b]	59 (55–63)	52 (43–60)	61 (57–66)	10 (1–19)
Fatigue Severity Scale, total score [a,b]	49 (46–51)	52 (48–56)	48 (44–51)	−4 (−10–2)
Fatigue VAS (0 to 100) [a,b]	44 (39–50)	43 (32–55)	45 (38–51)	2 (−11–14)
Neuro-QOL, t score [a,b]	41 (39–44)	39 (34–44)	42 (39–45)	3 (−3–9)
GAD-7, total score [a]	7 (6–8)	8 (5–10)	6 (5–8)	−1 (−4–1)
PHQ-2, total score [a]	2 (2–2)	2 (2–3)	2 (2–2)	0 (−1–0)
WHODAS, total score [a,b]	32 (28–37)	37 (29–45)	30 (25–36)	−7 (−17–4)

Data are presented as median (95% confidence interval (CI)) or [a] mean (95% CI). [b] EQ-5D-5L health status VAS, fatigue VAS: dysautonomia, n = 26; not classified, n = 70. Fatigue Severity Scale, GAD-7: dysautonomia, n = 26; not classified, n = 71; Neuro-QOL: dysautonomia, n = 16; not classified, n = 54; WHODAS: dysautonomia, n = 24; not classified, n = 65.

The majority of individuals who completed the physical activity survey questions reported a reduced engagement in moderate-intensity (69/85 (81%)) and vigorous-intensity (60/86 (70%)) physical activity post-acute COVID-19 infection. Both LVEDV (mean diff. (95% CI) −13 [−1–−26] mL, $p = 0.04$) and SV (−10 [−1–−20] mL, $p = 0.03$) measurements were smaller in those reporting reduced engagement in moderate-intensity physical activity post-COVID-19 infection when compared to those with similar or more regular engagement in moderate-intensity physical activity. Smaller LVMI measurements were weakly correlated with worse fatigue VAS scores ($r = 0.23$, $p = 0.02$).

4. Discussion

Both myocardial injury and the phenotype of Long COVID are important known consequences of infection with SARS-CoV-2. However, it is not yet established whether these phenomena are epidemiologically or pathobiologically related. Because Long COVID symptoms classically include some combination of fatigue, palpitations, breathlessness, chest discomfort and post-exertional symptom exacerbation, it is imperative to clarify whether these are manifestations of cardiac dysfunction. These data demonstrate that in the overwhelming majority of a set of well-characterized individuals evaluated in a Long COVID clinic, LV and RV function is normal. Of this cohort, only three individuals had evidence of new-onset cardiomyopathy (segmental or diffuse LV dysfunction and/or LV dilatation). Outside of these edge cases, the majority of individuals with Long COVID in this study did not require hospitalization for acute COVID-19, and it was found that their persistent symptoms were not associated with structural or functional cardiac abnormalities. These findings support that the phenomena of SARS-CoV-2-induced myocardial injury and Long COVID are not directly related. As such, a cardiac rehabilitation approach that might be appropriate following myocardial infarction or heart failure hospitalization is unlikely to be effective in treating Long COVID symptoms and should not be prescribed.

Previous data have shown that the typical Long COVID presentation incorporates features of dysautonomia, with individuals reporting systemic symptoms (fatigue, breathlessness, chest discomfort, palpitations, dizziness, syncope/presyncope, orthostatic exacerbations, leg pain and exercise intolerance/post-exertional symptom exacerbation). Further, the triggers of worsening symptoms (physical exertion, stress, dehydration, weather changes, consuming large meals, premenstrual period and alcohol) are similar [9]. Dysautonomia is now an established endotype of Long COVID [12–14]; yet, only 13–33% of individuals meet the criteria for specific diagnoses such as POTS [18–20]. It is, therefore, necessary to define a new terminology: post-COVID-19 dysautonomia (PCD).

The proposed pathophysiologic connections between COVID-19 and dysautonomia include (1) hypovolemia due to fever, decreased fluid intake, nausea, excessive diaphoresis and prolonged bed rest, leading to increased cardiac sympathetic nervous system outflow and cardiac atrophy; (2) direct SARS-CoV-2 infection and the destruction of extracardiac postganglionic sympathetic neurons, increasing sympathetic outflow; (3) SARS-CoV-2 invasion of the brainstem, resulting in increased central sympathetic outflow analogous to that seen in Takotsubo cardiomyopathy; and (4) virally induced autoimmunity, directing an immune attack against host neurons [23]. Further work is required to better understand which, if any, of these mechanisms underlie PCD.

Small cardiac size with reduced blood volume (i.e., cardiac atrophy) has previously been noted in POTS [17] and other forms of dysautonomia [16]. Here, it is reported that ~17% of individuals with Long COVID demonstrate features of small LV chamber size and that echocardiographic signs of cardiac atrophy correlate with reductions in moderate physical activity (lower LV EDV and SV) and worse fatigue (smaller LVMI). Clinicians should be attuned to recognizing PCD, as affected individuals may benefit from a number of effective interventions (i.e., oral fluid and electrolyte repletion as well as lower extremity compression garments), pharmacotherapy and/or autonomic rehabilitation [24]. The latter is a specialized form of rehabilitation that focuses on utilizing symptom-titrated exercises that focus on retraining appropriate physiological responses to autonomic challenges, and can improve or resolve symptoms [25,26] while increasing cardiac mass and blood volumes in a majority of cases [17,27]. Among the Long COVID cohort, only 22% of individuals were diagnosed with PCD, even though symptom type, severity and LV chamber size were not distinguishable from the unclassified group. Overall, significant overlap in clinical and cardiac profiles between the PCD and unclassified groups indicates that these represent a single phenotype, with the etiology of symptoms being appropriately recognized in only a fraction of LC patients. Only SV was lower in the PCD group, suggesting that there may have been features present in these individuals that raised suspicion for dysautonomia or prompted a work-up with formal autonomic testing. For example, individuals with lower SV may show physiologic signs such as sinus tachycardia or exaggerated exertional or orthostatic tachycardia.

For many years, dysautonomia (that has been triggered by conditions or events other than acute SARS-CoV-2 infection) has been associated with reduced cardiac size, and improvements in functional testing after rehabilitation have been associated with complementary increases in cardiac size [16,17]. This has led to the narrative that at least some dimension of the disability caused by dysautonomia may be caused by "cardiovascular deconditioning" [27]. However, few studies account for the fact that prior to the emergence of Long COVID, dysautonomia was not diagnosed until, on average, almost six years after the initial emergence of symptoms [28]. At this time, it is possible that significant deconditioning can occur as otherwise healthy and active individuals begin to avoid physical activity due to the propensity to produce post-exertional symptoms. With the emergence of Long COVID as a chronic post-acute infection syndrome and the observation that a large proportion of Long COVID cases are accompanied by dysautonomia [29], cases of PCD are being diagnosed much faster than dysautonomia that was diagnosed prior to 2020. In this study, there was a shorter period of time between the onset of PCD symptoms and its diagnosis (since all diagnoses of PCD in this study happened in under 2 years of symptom onset). This study's finding of minimal changes in cardiac size related to a diagnosis of PCD compared with those without a PCD diagnosis would indicate that such cardiac changes are in fact related to prolonged periods of inactivity, rather than any cardinal pathobiological features of dysautonomia. Few studies in the field have had the opportunity to study cardiac morphology of cases of dysautonomia that have been diagnosed so soon after the onset of symptoms.

The limitations of this study include the use of data collected retrospectively from a convenience sample and the lack of baseline information, such as pre-COVID-19 PROs and echocardiographic features for most individuals. The classification of dysautonomia by the research team relied on EHR information, which may result in issues such as variability between cardiologists and physicians in their assessment and diagnosis of dysautonomia, a lack of uniformity in the outcomes used to confirm a diagnosis dysautonomia including electrocardiograms and the potential for information to be absent from the EHR for both classified and unclassified groups.

The contemporary absence of universal diagnostic criteria for dysautonomia and, more specifically, PCD challenges the ability to identify individuals who would benefit from interventions. The improved recognition of PCD can address important disparities in healthcare, especially for women, who are more frequently affected. Although cohort-based observations support cardiac atrophy as a mechanism contributing to PCD, echocardiographic measurements are not likely to be useful as sole biomarkers to capture this phenotype. Further studies are needed to more rigorously define PCD and enable improved recognition and care.

5. Conclusions

The majority of individuals with Long COVID report shared symptoms and did not demonstrate cardiac dysfunction on echocardiography. Cardiac atrophy, as has been previously reported in association with other forms of dysautonomia, is a feature of Long COVID and correlates with reductions in physical activity levels and worse fatigue. Improved biomarkers of PCD are needed to enable better recognition and care for patients with Long COVID.

Author Contributions: Conceptualization, L.T., J.W., D.P. and A.K. (Amy Kontorovich); methodology, L.T., J.W., D.P. and A.K. (Amy Kontorovich).; software, L.T., J.W., D.P. and A.K. (Amy Kontorovich).; validation, L.T., J.W., D.P. and A.K. (Amy Kontorovich); formal analysis, J.W. and D.P.; investigation, L.T., J.W., D.P. and A.K. (Amy Kontorovich); resources, L.T., J.W., D.P. and A.K. (Amy Kontorovich); data curation, L.T., J.W., E.B., J.T.-M., A.K. (Amanda Kelly), K.W., C.Z., D.P. and A.K. (Amy Kontorovich); writing—original draft preparation, L.T., J.W., D.P. and A.K. (Amy Kontorovich); writing—review and editing, L.T., J.W., E.B., J.T.-M., A.K. (Amanda Kelly), K.W., C.Z., D.P. and A.K. (Amy Kontorovich); visualization, L.T., J.W., E.B., J.T.-M., A.K. (Amanda Kelly), K.W., C.Z., D.P. and A.K. (Amy Kontorovich); supervision, D.P. and A.K. (Amy Kontorovich); project administration, D.P. and A.K. (Amy Kontorovich). All authors have read and agreed to the published version of the manuscript.

Funding: This research received no external funding

Institutional Review Board Statement: Approval for publication was provided by the Mount Sinai Program for Protection of Human Subjects (IRB 21-00944).

Informed Consent Statement: A waiver of consent was approved by the Mount Sinai Program for Protection of Human Subjects (IRB 21-00944).

Data Availability Statement: Data supporting reported results can be provided upon reasonable request to the corresponding author.

Conflicts of Interest: The authors declare no conflict of interest.

References

1. Lala, A.; Johnson, K.W.; Januzzi, J.L.; Russak, A.J.; Paranjpe, I.; Richter, F.; Zhao, S.; Somani, S.; Van Vleck, T.; Vaid, A.; et al. Prevalence and Impact of Myocardial Injury in Patients Hospitalized With COVID-19 Infection. *J. Am. Coll Cardiol.* **2020**, *76*, 533–546. [CrossRef]
2. Nakatani, S.; Ohta-Ogo, K.; Nishio, M.; Amemiya, K.; Sato, S.; Sawano, H.; Hatakeyama, K.; Katano, H.; Suzuki, T.; Hirooka, K. Microthrombosis as a cause of fulminant myocarditis-like presentation with COVID-19 proven by endomyocardial biopsy. *Cardiovasc. Pathol.* **2022**, *60*, 107435. [CrossRef] [PubMed]
3. Farkouh, M.E.; Stone, G.W.; Lala, A.; Bagiella, E.; Moreno, P.R.; Nadkarni, G.N.; Ben-Yehuda, O.; Granada, J.F.; Dressler, O.; Tinuoye, E.O.; et al. Anticoagulation in Patients With COVID-19: JACC Review Topic of the Week. *J. Am. Coll. Cardiol.* **2022**, *79*, 917–928. [CrossRef] [PubMed]
4. Sperotto, F.; Friedman, K.G.; Son, M.B.F.; VanderPluym, C.J.; Newburger, J.W.; Dionne, A. Cardiac manifestations in SARS-CoV-2-associated multisystem inflammatory syndrome in children: A comprehensive review and proposed clinical approach. *Eur. J. Pediatr.* **2021**, *180*, 307–322. [CrossRef] [PubMed]
5. Aldeghaither, S.M.; Qutob, R.M.; Assanangkornchai, N.M.; Issa-Chergui, B.; Tam, M.C.; Larotondo, R.; Samoukovic, G.M. Clinical and Histopathologic Features of Myocarditis in Multisystem Inflammatory Syndrome (Adult)-Associated COVID-19. *Crit. Care Explor.* **2022**, *10*, e0630. [CrossRef] [PubMed]
6. Ammirati, E.; Lupi, L.; Palazzini, M.; Hendren, N.S.; Grodin, J.L.; Cannistraci, C.V.; Schmidt, M.; Hekimian, G.; Peretto, G.; Bochaton, T.; et al. Prevalence, Characteristics, and Outcomes of COVID-19-Associated Acute Myocarditis. *Circulation.* **2022**, *145*, 1123–1139. [CrossRef]
7. Moulson, N.; Petek, B.J.; Drezner, J.A.; Harmon, K.G.; Kliethermes, S.A.; Patel, M.R.; Baggish, A.L.; Asif, I.M.; Borchers, J.; Edenfield, K.M.; et al. SARS-CoV-2 Cardiac Involvement in Young Competitive Athletes. *Circulation.* **2021**, *144*, 256–266. [CrossRef] [PubMed]
8. Daniels, C.J.; Rajpal, S.; Greenshields, J.T.; Rosenthal, G.L.; Chung, E.H.; Terrin, M.; Jeudy, J.; Mattson, S.E.; Law, I.H.; Borchers, J.; et al. Prevalence of Clinical and Subclinical Myocarditis in Competitive Athletes With Recent SARS-CoV-2 Infection: Results From the Big Ten COVID-19 Cardiac Registry. *JAMA Cardiol.* **2021**, *6*, 1078–1087. [CrossRef]
9. Tabacof, L.; Tosto-Mancuso, J.; Wood, J.; Cortes, M.; Kontorovich, A.; McCarthy, D.; Rizk, D.; Rozanski, G.; Breyman, E.; Nasr, L.; et al. Post-acute COVID-19 Syndrome Negatively Impacts Physical Function, Cognitive Function, Health-Related Quality of Life, and Participation. *Am. J. Phys. Med. Rehabil.* **2022**, *101*, 48–52. [CrossRef]
10. Davis, H.E.; Assaf, G.S.; McCorkell, L.; Wei, H.; Low, R.J.; Re'Em, Y.; Redfield, S.; Austin, J.P.; Akrami, A. Characterizing long COVID in an international cohort: 7 months of symptoms and their impact. *EClinicalMedicine* **2021**, *38*, 101019. [CrossRef]
11. Centers for Disease Control and Prevention. Nearly One in Five American Adults Who Have Had COVID-19 Still Have "Long COVID". Available online: https://www.cdc.gov/nchs/pressroom/nchs_press_releases/2022/20220622.htm# (accessed on 25 July 2022).
12. Blitshteyn, S.; Whitelaw, S. Postural orthostatic tachycardia syndrome (POTS) and other autonomic disorders after COVID-19 infection: A case series of 20 patients. *Immunol. Res.* **2021**, *69*, 205–211. [CrossRef] [PubMed]
13. Johansson, M.; Ståhlberg, M.; Runold, M.; Nygren-Bonnier, M.; Nilsson, J.; Olshansky, B.; Bruchfeld, J.; Fedorowski, A. Long-Haul Post-COVID-19 Symptoms Presenting as a Variant of Postural Orthostatic Tachycardia Syndrome: The Swedish Experience. *JACC Case Rep.* **2021**, *3*, 573–580. [CrossRef] [PubMed]
14. Buoite Stella, A.; Furlanis, G.; Frezza, N.A.; Valentinotti, R.; Ajcevic, M.; Manganotti, P. Autonomic dysfunction in post-COVID patients with and witfhout neurological symptoms: A prospective multidomain observational study. *J. Neurol.* **2022**, *269*, 587–596. [CrossRef]
15. Jamal, S.M.; Landers, D.B.; Hollenberg, S.M.; Turi, Z.G.; Glotzer, T.V.; Tancredi, J.; Parrillo, J.E. Prospective Evaluation of Autonomic Dysfunction in Post-Acute Sequela of COVID-19. *J. Am. Coll. Cardiol.* **2022**, *79*, 2325–2330. [CrossRef] [PubMed]
16. Ruiz Maya, T.; Fettig, V.; Mehta, L.; Gelb, B.D.; Kontorovich, A.R. Dysautonomia in hypermobile Ehlers-Danlos syndrome and hypermobility spectrum disorders is associated with exercise intolerance and cardiac atrophy. *Am. J. Med. Genet. A.* **2021**, *185*, 3754–3761. [CrossRef]
17. Fu, Q.; VanGundy, T.B.; Galbreath, M.M.; Shibata, S.; Jain, M.; Hastings, J.L.; Bhella, P.S.; Levine, B.D. Cardiac origins of the postural orthostatic tachycardia syndrome. *J. Am. Coll. Cardiol.* **2010**, *55*, 2858–2868. [CrossRef]

18. Novak, P.; Mukerji, S.S.; Alabsi, H.S.; Systrom, D.; Marciano, S.P.; Felsenstein, D.; Mullally, W.J.; Pilgrim, D.M. Multisystem Involvement in Post-Acute Sequelae of Coronavirus Disease 19. *Ann. Neurol.* **2022**, *91*, 367–379. [CrossRef]
19. Shouman, K.; Vanichkachorn, G.; Cheshire, W.P.; Suarez, M.D.; Shelly, S.; Lamotte, G.J.; Sandroni, P.; Benarroch, E.E.; Berini, S.E.; Cutsforth-Gregory, J.K.; et al. Autonomic dysfunction following COVID-19 infection: An early experience. *Clin. Auton. Res.* **2021**, *31*, 385–394. [CrossRef]
20. Monaghan, A.; Jennings, G.; Xue, F.; Byrne, L.; Duggan, E.; Romero-Ortuno, R. Orthostatic Intolerance in Adults Reporting Long COVID Symptoms Was Not Associated With Postural Orthostatic Tachycardia Syndrome. *Front. Physiol.* **2022**, *13*, 833650. [CrossRef]
21. Centers for Disease Control and Prevention. Post-COVID Conditions: Overview for Healthcare Providers. Available online: https://www.cdc.gov/coronavirus/2019-ncov/hcp/clinical-care/post-covid-conditions.html (accessed on 25 July 2022).
22. Lang, R.M.; Badano, L.P.; Mor-Avi, V.; Afilalo, J.; Armstrong, A.; Ernande, L.; Flachskampf, F.A.; Foster, E.; Goldstein, S.A.; Kuznetsova, T.; et al. Recommendations for cardiac chamber quantification by echocardiography in adults: An update from the American Society of Echocardiography and the European Association of Cardiovascular Imaging. *J. Am. Soc. Echocardiogr.* **2015**, *28*, 1–39.e14. [CrossRef]
23. Goldstein, D.S. The possible association between COVID-19 and postural tachycardia syndrome. *Heart Rhythm.* **2021**, *18*, 508–509. [CrossRef]
24. Bryarly, M.; Phillips, L.T.; Fu, Q.; Vernino, S.; Levine, B.D. Postural Orthostatic Tachycardia Syndrome: JACC Focus Seminar. *J. Am. Coll. Cardiol.* **2019**, *73*, 1207–1228. [CrossRef] [PubMed]
25. George, S.A.; Bivens, T.B.; Howden, E.J.; Saleem, Y.; Galbreath, M.M.; Hendrickson, D.; Fu, Q.; Levine, B.D. The international POTS registry: Evaluating the efficacy of an exercise training intervention in a community setting. *Heart Rhythm.* **2016**, *13*, 943–950. [CrossRef] [PubMed]
26. Winker, R.; Barth, A.; Bidmon, D.; Ponocny, I.; Weber, M.; Mayr, O.; Robertson, D.; Diedrich, A.; Maier, R.; Pilger, A.; et al. Endurance exercise training in orthostatic intolerance: A randomized, controlled trial. *Hypertension.* **2005**, *45*, 391–398. [CrossRef] [PubMed]
27. Fu, Q.; Vangundy, T.B.; Shibata, S.; Auchus, R.J.; Williams, G.H.; Levine, B.D. Exercise training versus propranolol in the treatment of the postural orthostatic tachycardia syndrome. *Hypertension* **2011**, *58*, 167–175. [CrossRef] [PubMed]
28. Pavlik, D.; Agnew, D.; Stiles, L.; Ditoro, R. Recognizing postural orthostatic tachycardia syndrome. *JAAPA* **2016**, *29*, 17–23. [CrossRef]
29. Blitshteyn, S.; Whiteson, J.H.; Abramoff, B.; Azola, A.; Bartels, M.N.; Bhavaraju-Sanka, R.; Chung, T.; Fleming, T.K.; Henning, E.; Miglis, M.G.; et al. Multi-disciplinary collaborative consensus guidance statement on the assessment and treatment of autonomic dysfunction in patients with post-acute sequelae of SARS-CoV-2 infection (PASC). *PMR* **2022**, *14*, 1270–1291.

Journal of
Personalized
Medicine

MDPI

Opinion

Possible Role of Fibrinaloid Microclots in Postural Orthostatic Tachycardia Syndrome (POTS): Focus on Long COVID

Douglas B. Kell [1,2,3,*], Muhammed Asad Khan [4], Binita Kane [1,5], Gregory Y. H. Lip [6,7] and Etheresia Pretorius [1,3,*]

1 Department of Biochemistry, Cell and Systems Biology, Institute of Systems, Molecular and Integrative Biology, Faculty of Health and Life Sciences, University of Liverpool, Crown St, Liverpool L69 7ZB, UK; binita.kane@mft.nhs.uk
2 The Novo Nordisk Foundation Centre for Biosustainability, Building 220, Chemitorvet 200, Technical University of Denmark, 2800 Kongens Lyngby, Denmark
3 Department of Physiological Sciences, Faculty of Science, Stellenbosch University, Stellenbosch Private Bag X1, Matieland 7602, South Africa
4 Directorate of Respiratory Medicine, Manchester University Hospitals, Wythenshawe Hospital, Manchester M23 9LT, UK; asad.khan12@nhs.net
5 Manchester University Foundation Trust and School of Biological Sciences, University of Manchester, Manchester M13 9PL, UK
6 Liverpool Centre for Cardiovascular Science at University of Liverpool, Liverpool John Moores University and Liverpool Heart & Chest Hospital, Liverpool L14 3PE, UK; gregory.lip@liverpool.ac.uk
7 Danish Center for Health Services Research, Department of Clinical Medicine, Aalborg University, 9220 Aalborg, Denmark
* Correspondence: dbk@liv.ac.uk (D.B.K.); resiap@sun.ac.za (E.P.)

Abstract: Postural orthostatic tachycardia syndrome (POTS) is a common accompaniment of a variety of chronic, inflammatory diseases, including long COVID, as are small, insoluble, 'fibrinaloid' microclots. We here develop the argument, with accompanying evidence, that fibrinaloid microclots, through their ability to block the flow of blood through microcapillaries and thus cause tissue hypoxia, are not simply correlated with but in fact, by preceding it, may be a chief intermediary cause of POTS, in which tachycardia is simply the body's exaggerated 'physiological' response to hypoxia. Similar reasoning accounts for the symptoms bundled under the term 'fatigue'. Amyloids are known to be membrane disruptors, and when their targets are nerve membranes, this can explain neurotoxicity and hence the autonomic nervous system dysfunction that contributes to POTS. Taken together as a system view, we indicate that fibrinaloid microclots can serve to link POTS and fatigue in long COVID in a manner that is at once both mechanistic and explanatory. This has clear implications for the treatment of such diseases.

Keywords: fibrinaloid microclots; postural orthostatic tachycardia syndrome (POTS); Long COVID; TeamClots

1. Introduction

Orthostasis, Orthostatic Intolerance, and POTS

Human beings have evolved to maintain a largely erect posture [1] and can adopt it from recumbent poses. Orthostasis describes the (normal) physiological response used to counteract the potential fall in blood pressure when a person who has been lying down assumes the upright position. This tendency occurs because, in an adult, gravity causes a shift of some 300 to 800 mL of blood from the upper to the lower body. This orthostasis depends strongly on the autonomic nervous system.

However, if the system does not respond properly, there can be a significant decrease in the central blood pressure; common symptoms of such hypoperfusion are dizziness, lightheadedness, and syncope (fainting). The resulting intolerance of the upright posture is known as orthostatic intolerance (OI). When accompanied by a sustained postural drop in blood pressure (of more than 20 mmHg systolic or 10 mmHg diastolic [2]), the patient is diagnosed with orthostatic hypotension, which is a form of orthostatic intolerance (OI). Another variant of OI occurs when there is less of a fall in blood pressure, but the autonomic response leads instead to a rapid increase in heart rate (tachycardia). This is known as postural orthostatic tachycardia syndrome (POTS) (e.g., [3–5]). POTS is a manifestation of autonomic dysregulation and is clinically characterized as excessive tachycardia upon standing in the presence of symptomatic orthostatic intolerance. We recognize that POTS may be classified into subtypes such as neuropathic POTS and hyperadrenergic POTS; however, most of the papers we cite do not in fact make this distinction, and, for the present purposes, we avoid doing so as well, since our chief aim here was simply to suggest that there is, in general, significant evidence for the role of fibrinaloid microclots in POTS.

Although well known in other contexts for at least three decades [6,7] (see Table 1), with at least 500,000 cases in the USA alone [8–10], mostly in women (5:1) [5,9,11–14], POTS has emerged as a frequent symptom of both acute [15] and long COVID (e.g., [16–21] as part of the wider cardiovascular dysautonomia spectrum; see Table 1).

Table 1. Some diseases and syndromes with which POTS is associated.

Disease, State, or Syndrome	Comments	Selected Reference(s)
Autoimmune disorders and Autoimmunity	Some strong associations	[16,22–26]
Cognitive function	Large amount of literature; improved by plasma exchange [27]	[27–30]
Fatigue		[31–38]
HPV or other antiviral vaccination	An example of induction by a viral protein	[39–45] but cf. [46]
Inflammation		[47]
Irritable bowel disease		[48]
Long COVID	A very common occurrence and a focus of our interest	[16–20,49–68]
Migraine		[69]
Multiple sclerosis	Now recognized as possibly caused by Epstein–Barr virus [70] (albeit much earlier evidence for an infectious origin existed [71,72], cf. [73,74]).	[75]
Myalgic encephalomyelitis/chronic fatigue syndrome (ME/CFS)	Is also usually a postviral disease and bears a number of similarities to long COVID [68,76–79]	[31,32,52,80–84]
Platelet delta granule storage pool deficiency	Causal direction unclear	[85]
Pregnancy	Many cardiovascular stresses accompany pregnancy, especially during hypertensive disorders [86,87]	[88,89]
Reviews		[22]

The management of POTS has been the subject of prior reviews and guidelines and is beyond the aims of the present study [90,91]. Our focus in this study was mainly on microclots as a plausible, mechanistic basis for POTS, especially in relation to long COVID.

2. The Normal Control of Heart Rate

Because of the general interest in POTS in long COVID and other affected communities, we include a very brief and high-level overview. The heart rate is controlled by many genetic and lifestyle factors (e.g., [92,93]), and the required kinds of understanding are both conceptual (e.g., the need to cater for the time-varying demands of tissue oxygenation) and mechanistic (e.g., the involvement of the endocrine and autonomic nervous systems). Our overview here is very far from being comprehensive, and our focus is necessarily on short-term control, where the autonomic nervous system is predominant (Figure 1, after [92]).

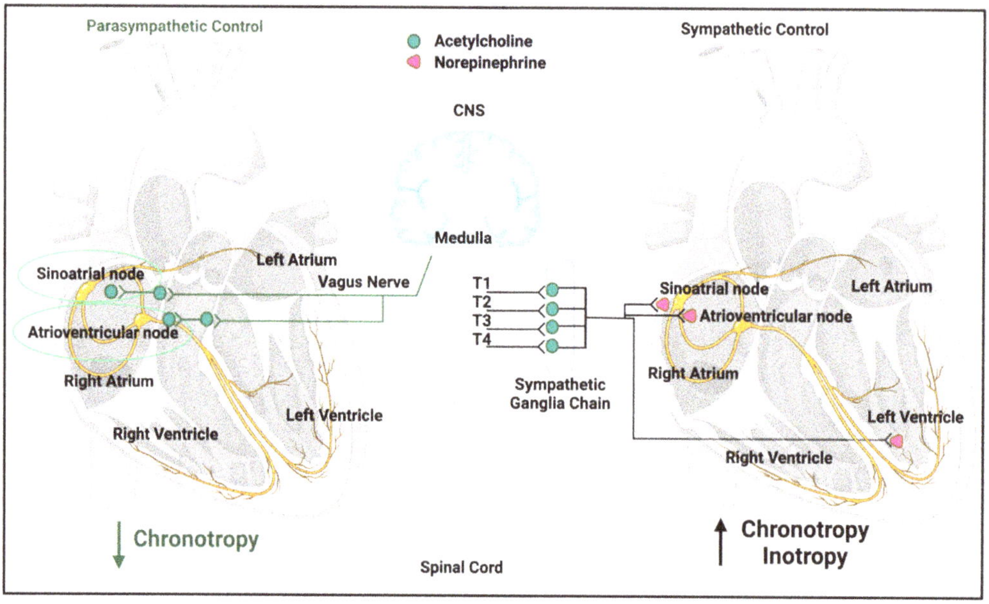

Figure 1. Autonomic nervous system regulation of heart function (after [92]). Created with BioRender.com. Access date: 26 November 2023.

As summarized in Figure 1 (redrawn from [92]), both the sympathetic and parasympathetic branches of the autonomic nervous system are involved. The former is more involved in stress responses (often called 'fight-or-flight') and can release noradrenaline (norepinephrine) to increase heart rate, whilst the latter (often called 'rest-and-digest') underpins basal activity via the vagus nerve that can release acetylcholine to decrease heart rate relative to its base rate. Multiple control steps involve baroceptors that sense pressure and other receptors that respond to pH, hypoxia, and hypercapnia. In particular, under most conditions, the heart necessarily and appropriately responds to acute hypoxia by increasing heart rate (e.g., [94–98]).

3. Diagnosis of POTS

Most chronic, inflammatory diseases—as their name suggests—possess multiple common symptoms [99], while those such as long COVID characterized by subsets of multiple symptoms can easily be subclustered (e.g., [49,100–102]). The earlier definition of POTS comes from a very small study of 16 patients in 1993, of whom, interestingly, 7 were thought to have had previous viral infections [6,103]. Nowadays, for instance, the Canadian Cardiology Society has published a position paper describing a wider heterogenous range of clinical syndromes and a spectrum of orthostatic intolerance; they propose that discrete subtypes are identified over time, each with different underlying pathophysiological phenotypes that allow for specific targeted treatment [90]. However, for present purposes, in the case of POTS, both the high-level definition and the diagnosis are relatively straightforward, as they follow virtually from the name: heart rate is monitored for tachycardia (an increase in heart rate exceeding 30 beats per minute (bpm) within the initial 10 min of standing or head-up tilt (HUT)- or a 'final' value exceeding 120 bpm) as the individual changes their posture from horizontal to (more) vertical [5].

Differences can occur because the transition is commonly affected either by active standing or a passive 'tilt table' test [104–107]. The latter, which is somewhat more controlled and considered more reliable [108], commonly involves a 'head-up tilt' in which an individual is strapped to a horizonal table and commonly tilted to an angle of 60–80° [106,109], and heart rate and other measurements are performed. Transcranial doppler ultrasound may be used to detect blood flow [110]. It is recognized that such 'provocative' tests are of most value when individuals record similar symptoms to those that they normally experience [111]. For all events, the conceptual recognition of POTS is to be seen as reasonably straightforward [112,113]. It is important to recognize that the diagnostic criteria for heart rate changes are arbitrary and based on small case series, and that patients can have disabling OI and other symptoms of autonomic dysfunction without meeting the traditional cutoffs; this is no different in long COVID patients presenting with symptoms of POTS.

4. Occurrence and Comorbidities of POTS

Although we did not cover POTS (nor even autonomic dysfunction) in our earlier review of chronic, inflammatory diseases [99], the occurrence of POTS, which is highly heterogeneous [114], broadly mirrors the kinds of disease that we did mention there. Table 1 lists some of them, implying elements of a common origin. Of particular interest is the evidence for endothelial microvascular dysfunction [50], which can occur via the microclot-mediated blockage of red cell flow to tissues.

5. Dysautonomia

Autonomic dysfunction (dysautonomia) describes any malfunction in the autonomic nervous system, especially the vagus nerve [115,116], which is a key element in (but not synonymous with [117]) POTS, and the occurrence of dysautonomia broadly mirrors the diseases in which POTS is known to occur (Table 2).

Table 2. Some diseases and syndromes in which dysautonomia is known to occur.

Disease, State, or Syndrome	Comments	Selected Reference(s)
Familial (monogenic)	Lesion in the IKBKAP gene	[118]
Long COVID		[57,60,62,63,67,76,119–124]
Multiple sclerosis		[125,126]
Myalgic encephalomyelitis/chronic fatigue syndrome		[76,82,119,127–133]
Parkinson's disease		[134]

6. Fatigue and POTS

Like POTS, fatigue is a common accompaniment of many acute and chronic inflammatory diseases. It is usually based on scoring questionnaires and thus lacks a crisp definition [135–142]. However, fatigue is generally used to cover a debilitating set of symptoms in which attempts to carry out what would normally be considered a very mild exertion are followed immediately by an inability to perform or to continue such exertions and a period in which extreme rest is required. In contrast to physiological 'tiredness', rest and sleep are not physically or mentally rejuvenating in fatigue. As noted in Table 1 [31–37], fatigue is a common accompaniment of POTS and—as we shall argue—likely has a main common cause.

7. The Role of Fibrinaloid Microclots in POTS

Although the origins of our discoveries that blood could clot into a very anomalous form lie earlier- in observations using the electron microscope (e.g., [143–146])- it was not until 2016 [147] that we determined using fluorescence microscopy that these anomalous forms were in fact amyloid in nature [148–152], that they could be induced by highly substoichiometric amounts of bacterial lipopolysaccharide [147], and that the electron and optical microscopies were congruent [153]. Essentially all the clots visible using fluorescence staining were those visible in the bright field [154,155]. The microclots were found to be particularly prevalent in diabetes [156–158] and in particular in both acute [158] and long COVID [159–166], where they could be induced by miniscule concentrations of the spike protein [167,168]. They were also much raised over those in controls in individuals with ME/CFS [169,170]. Note that the generation of fibrinaloid microclots is essentially instantaneous (on the timescale of normal clotting) (e.g., [147,167]), whereas the time taken to develop POTS is slower. This is at least consistent with a causative role of the earlier-appearing microclots in the generation of the later-appearing POTS.

Microclots differ from clots mostly by being considerably smaller (broadly in the range of 1–200 µm, mostly at the lower end) (see Figure 2) and by virtue both of the adoption of an amyloid form [148,159,161] and their entrapment of molecules such as α_2-antiplasmin [163]. These and other properties [171] make them particularly resistant to fibrinolysis, so they are removed far less quickly than would normally be the case.

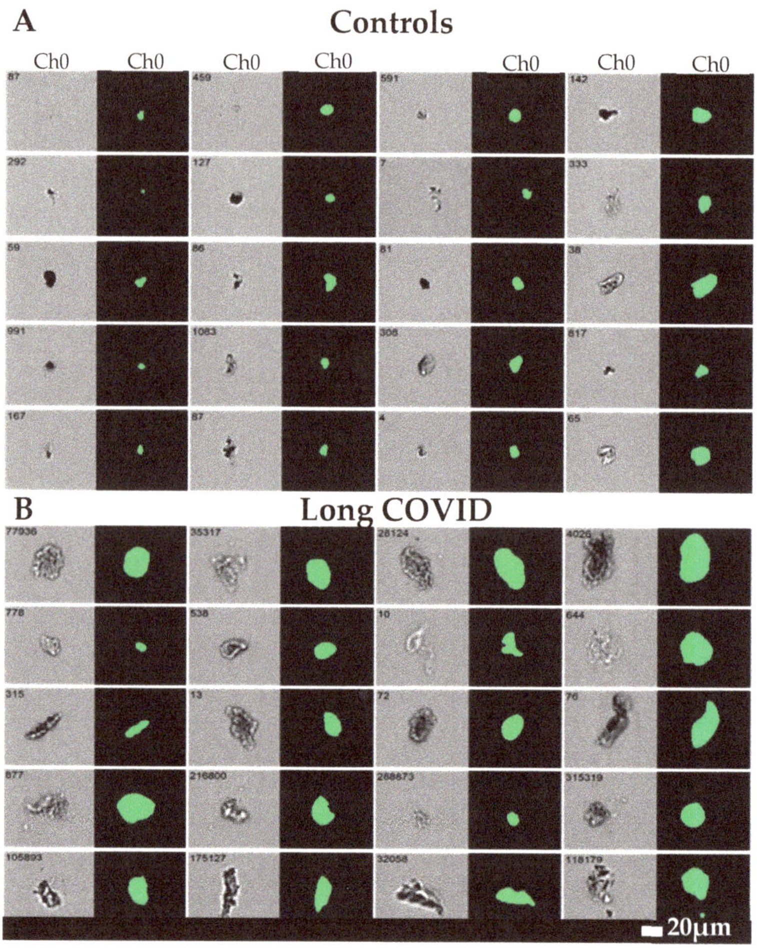

Figure 2. Microclot size distribution as seen with imaging flow cytometry (taken from [166]). Representative micrographs of microclots in (**A**) controls and (**B**) long COVID patients using an imaging flow cytometer. The brightfield images are displayed in Channel 1 (Ch01) and fluorescence intensity due to ThT binding in Channel 7 (Ch07). All images were captured using a 20x objective. The event number is displayed in the top-left corner of each image. NB: In these pictures, the POTS status of the individuals was not assessed.

A straightforward consequence of these insoluble fibrinaloid microclots is that as blood flow pushes them along, they can block up microcapillaries, thereby inhibiting the flux of oxygen-carrying red blood cells and thus inducing tissue hypoxia. Sensing low tissue oxygen concentrations naturally (as when exercising) may induce tachycardia, and this would provide a very ready explanation of both POTS and the fatigue that is a common occurrence in both ME/CFS and long COVID (see Figure 3).

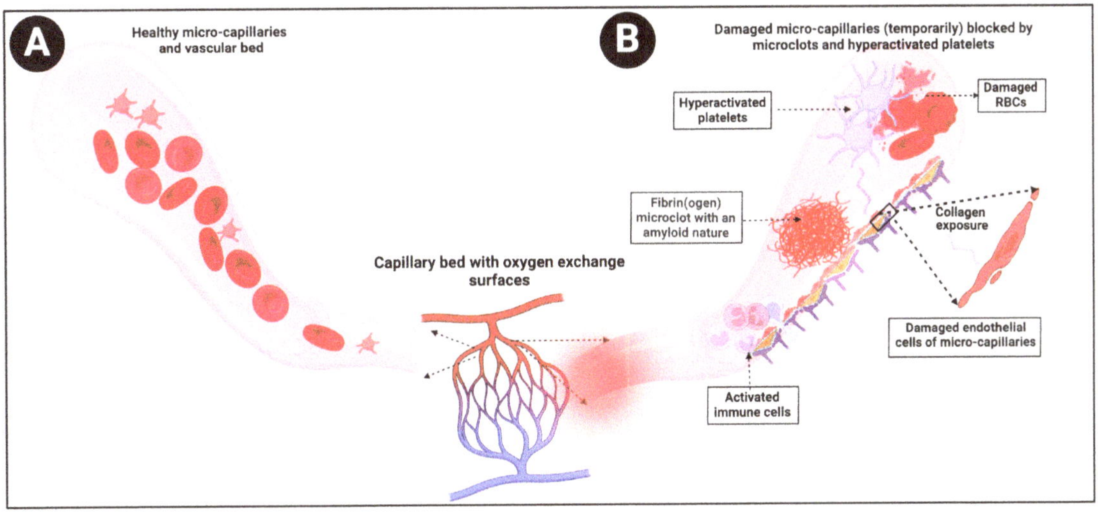

Figure 3. (**A**) Representation of healthy blood flow in microcapillaries (**B**) versus in an individual where damaged microcapillaries are (temporarily) blocked by microclots. Created with BioRender.com (accessed on 26 November 2023).

Other mechanisms for POTS in long COVID may include:

1. Relative hypovolemia secondary to inadequate peripheral vasoconstriction. This results in a reduction in stroke volume and cardiac output, causing the inhibition of tissue oxygen supply and the consequent compensatory tachycardia.
2. Small fiber neuropathy (SFN) has been well described in long COVID (e.g., [63,65,68,172]) and is a recognized cause of dysautonomia in the condition. SFN in long COVID can be driven by autoantibodies (already known to be associated with POTS and OH) or, potentially, by ischemia of the small fibres due to microclots.

8. The Role of Microclots in Fatigue

Just as the blocking of microcapillaries by microclots gives a ready explanation for POTS, it also gives a ready explanation for fatigue as tissues that rely on aerobic respiration for their normal function are deprived of oxygen. Specifically, the microclots vary widely in diameter, so they can migrate to those parts of the capillary bed where they can block the flow of red blood cells most effectively. Consequently, the affected tissues simply cannot perform their normal functions. While details vary for every individual, the existence and capillary-blocking behavior of the microclots also provide a simple and mechanistic explanation for the co-occurrence [31–33,35–37] of POTS and fatigue.

9. Relationship between Dysautonomia and Microclots

We know that molecules such as LPS (e.g., [147,149,150]) and the spike protein of SARS-CoV-2 (e.g., [154,158,159,163–167,173]) can cause microclots, such that any damage such molecules may cause to nerves may be indirect [174–176]. This said, it is reasonable that any damage to the membranes of nerves might be mediated via fibrinaloid microclots.

To this end, although the direct experiments have not been performed with fibrinaloid microclots (nor is it easy to conduct them in vivo), it is at least worth repeating that it is well established that amyloid forms of proteins (including those binding cations [177]) generally can effect damage to all kinds of phospholipid membranes directly (e.g., [177–202]). A variety of mechanisms have been proposed, such as those in Figure 4 [201].

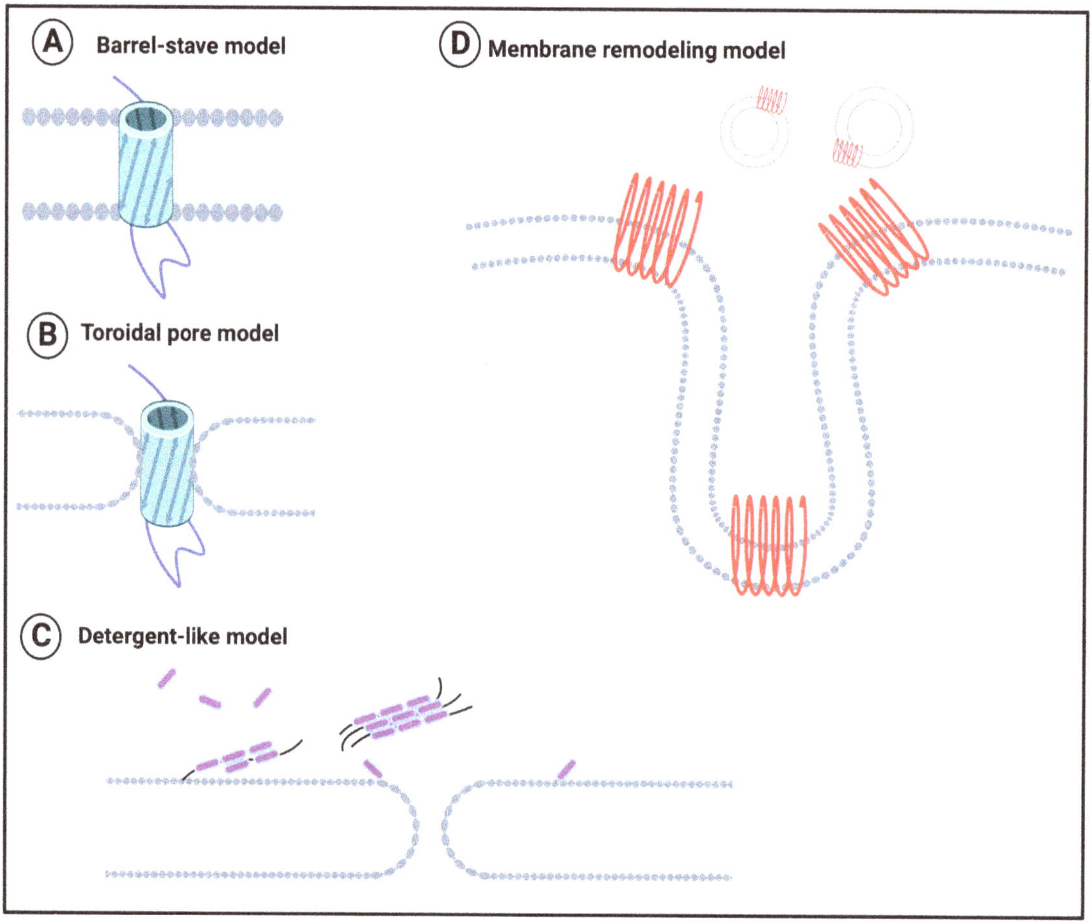

Figure 4. Membrane disruption models (redrawn from [201]). (**A**) The barrel-stave model suggests that proteins perpen-dicularly insert into the phospholipid bilayer plane, with the hydrophobic regions of protein oligomers contacting the hydrophobic interior of the membrane. (**B**) The toroidal pore model suggests that proteins insert perpendicular to the phospholipid bilayer, with the protein hydrophilic ends remaining in contact with the lipid head layer. (**C**) The deter-gent-like model, suggests that positively charged residues in the amyloidogenic protein bind to the membrane. (**D**) The membrane remodeling model suggests that membrane-bound peptides self-assemble into β-sheets that subsequently either form pores on the membrane surface (Pore formation model) or drag lipids out of the bilayer core (Detergent-like model). Created with BioRender.com (accessed on 26 November 2023).

When the membrane in question is a nerve membrane, neurotoxicity (e.g., [198,203–209] (leading to autonomic nervous system dysfunction) may result.

10. Systems Overview and Conclusions

We established that fibrinaloid microclots accompany a variety of diseases in which POTS is frequently diagnosed, with fatigue as a frequent feature, as are autoantibodies [161], implying a similar kind of cause or at least intermediate. The microclots do seem to fulfill this intermediary role, as they also provide a realistic set of mechanisms. This said, it should be admitted that detailed temporal studies have not been conducted in animals (which may not even provide a decent model), while those studies that did test, e.g., SARS-CoV-2 infection, in human volunteers directly [210] did not seek to measure microclots.

Very recently, Wüst and colleagues showed a variety of defects in the skeletal muscle of long COVID patients, including both amyloid deposition and mitochondrial dysfunction [211]. Coupled with the evidence for lactate overproduction in both COVID-19 [212–217] and ME/CFS [133,218–222], both of which are associated with POTS (Table 1), this provides further evidence for a role of inadequate O_2 uptake in these processes.

The system biology diagram linking these high-level elements is given in Figure 5.

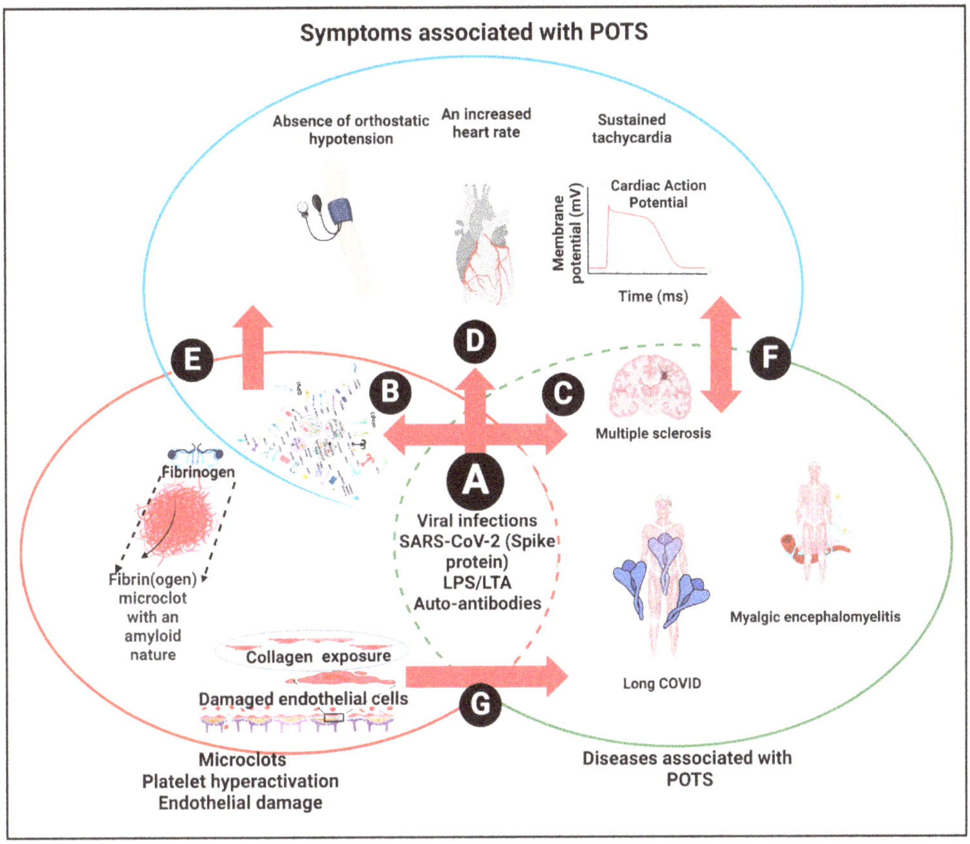

Figure 5. A system approach to defining dysautonomia. (A) Various causes of disease and symptoms resulting in vascular damage, microclots, and platelet hyperactivation (B) known to be involved in a variety of diseases (C) and in POTS (D). Similarly, vascular damage pathologies cause POTS (E) and other diseases (F), while POTS is found in various diseases (G). Created with BioRender.com (accessed on 26 November 2023).

We conclude that the presence of fibrinaloid microclots can indeed significantly account for the symptoms of POTS associated with long COVID (and likely other syndromes), just as they can for other symptoms [159], post-exertional symptom exacerbation [160], and the generation of autoantibodies [161].

Author Contributions: Conceptualization, D.B.K. and E.P.; writing—original draft preparation, D.B.K.; writing—review and editing, All Authors; visualization, E.P.; funding acquisition, E.P. & D.B.K. All authors have read and agreed to the published version of the manuscript.

Funding: EP: Funding was provided by the NRF of South Africa (grant number 142142), SA MRC (self-initiated research (SIR) grant), and the Balvi Foundation (grant B31). DBK thanks the Balvi Foundation (grant 18) and the Novo Nordisk Foundation for funding (grant NNF10CC1016517). The content and findings reported and illustrated are the sole deduction, view, and responsibility of the researchers and do not reflect the official position and sentiments of the funders.

Conflicts of Interest: E.P. is a named inventor on a patent application covering the use of fluorescence methods for microclot detection in long COVID.

References

1. Niemitz, C. The evolution of the upright posture and gait—A review and a new synthesis. *Naturwissenschaften* **2010**, *97*, 241–263. [CrossRef]
2. Freeman, R.; Wieling, W.; Axelrod, F.B.; Benditt, D.G.; Benarroch, E.; Biaggioni, I.; Cheshire, W.P.; Chelimsky, T.; Cortelli, P.; Gibbons, C.H.; et al. Consensus statement on the definition of orthostatic hypotension, neurally mediated syncope and the postural tachycardia syndrome. *Auton. Neurosci.* **2011**, *161*, 46–48. [CrossRef]
3. Fedorowski, A. Postural orthostatic tachycardia syndrome: Clinical presentation, aetiology and management. *J. Intern. Med.* **2019**, *285*, 352–366. [CrossRef]
4. Vernino, S.; Bourne, K.M.; Stiles, L.E.; Grubb, B.P.; Fedorowski, A.; Stewart, J.M.; Arnold, A.C.; Pace, L.A.; Axelsson, J.; Boris, J.R.; et al. Postural orthostatic tachycardia syndrome (POTS): State of the science and clinical care from a 2019 national institutes of health expert consensus meeting—Part 1. *Auton. Neurosci.* **2021**, *235*, 102828. [CrossRef]
5. Grubb, A.F.; Grubb, B.P. Postural orthostatic tachycardia syndrome: New concepts in pathophysiology and management. *Trends Cardiovasc. Med.* **2023**, *33*, 65–69. [CrossRef]
6. Schondorf, R.; Low, P.A. Idiopathic postural orthostatic tachycardia syndrome: An attenuated form of acute pandysautonomia? *Neurology* **1993**, *43*, 132–137. [CrossRef] [PubMed]
7. Safavi-Naeini, P.; Razavi, M. Postural orthostatic tachycardia syndrome. *Tex. Heart Inst. J.* **2020**, *47*, 57–59. [CrossRef] [PubMed]
8. Bryarly, M.; Phillips, L.T.; Fu, Q.; Vernino, S.; Levine, B.D. Postural orthostatic tachycardia syndrome: JACC focus seminar. *J. Am. Coll. Cardiol.* **2019**, *73*, 1207–1228. [CrossRef] [PubMed]
9. Shaw, B.H.; Stiles, L.E.; Bourne, K.; Green, E.A.; Shibao, C.A.; Okamoto, L.E.; Garland, E.M.; Gamboa, A.; Diedrich, A.; Raj, V.; et al. The face of postural tachycardia syndrome—Insights from a large cross-sectional online community-based survey. *J. Intern. Med.* **2019**, *286*, 438–448. [CrossRef] [PubMed]
10. Lecheler, L.; Hoffmann, F.; Tank, J.; Jordan, J. Run vagus run: Cardiovagal baroreflex function and the postural tachycardia syndrome. *Hypertension* **2021**, *77*, 1245–1247. [CrossRef] [PubMed]
11. Agarwal, A.K.; Garg, R.; Ritch, A.; Sarkar, P. Postural orthostatic tachycardia syndrome. *Postgrad. Med. J.* **2007**, *83*, 478–480. [CrossRef] [PubMed]
12. Low, P.A.; Sandroni, P.; Joyner, M.; Shen, W.K. Postural tachycardia syndrome (POTS). *J. Cardiovasc. Electrophysiol.* **2009**, *20*, 352–358. [CrossRef] [PubMed]
13. Hamrefors, V.; Spahic, J.M.; Nilsson, D.; Senneby, M.; Sutton, R.; Melander, O.; Fedorowski, A. Syndromes of orthostatic intolerance and syncope in young adults. *Open Heart* **2017**, *4*, e000585. [CrossRef] [PubMed]
14. Arnold, A.C.; Ng, J.; Raj, S.R. Postural tachycardia syndrome—Diagnosis, physiology, and prognosis. *Auton. Neurosci.* **2018**, *215*, 3–11. [CrossRef] [PubMed]
15. Hassani, M.; Fathi Jouzdani, A.; Motarjem, S.; Ranjbar, A.; Khansari, N. How COVID-19 can cause autonomic dysfunctions and postural orthostatic syndrome? A review of mechanisms and evidence. *Neurol. Clin. Neurosci.* **2021**, *9*, 434–442. [CrossRef]
16. El-Rhermoul, F.Z.; Fedorowski, A.; Eardley, P.; Taraborrelli, P.; Panagopoulos, D.; Sutton, R.; Lim, P.B.; Dani, M. Autoimmunity in long COVID and POTS. *Oxf. Open Immunol.* **2023**, *4*, iqad002. [CrossRef]
17. Blitshteyn, S.; Whiteson, J.; Abramoff, B.A.; Azola, A.; Bartels, M.N.; Bhavaraju-Sanka, R.; Chung, T.; Fleming, T.K.; Henning, E.; Miglis, M.G.; et al. Multi-disciplinary collaborative consensus guidance statement on the assessment and treatment of autonomic dysfunction in patients with post-acute sequelae of SARS-CoV-2 infection (PASC). *PM&R* **2022**, *14*, 1270–1291. [CrossRef]
18. Mallick, D.; Goyal, L.; Chourasia, P.; Zapata, M.R.; Yashi, K.; Surani, S. COVID-19 induced postural orthostatic tachycardia syndrome (POTS): A review. *Cureus* **2023**, *15*, e36955. [CrossRef] [PubMed]
19. Narasimhan, B.; Calambur, A.; Moras, E.; Wu, L.; Aronow, W. Postural orthostatic tachycardia syndrome in COVID-19: A contemporary review of mechanisms, clinical course and management. *Vasc. Health Risk Manag.* **2023**, *19*, 303–316. [CrossRef]

20. Pollack, B.; von Saltza, E.; McCorkell, L.; Santos, L.; Hultman, A.; Cohen, A.K.; Soares, L. Female reproductive health impacts of long COVID and associated illnesses including ME/CFS, POTS, and connective tissue disorders: A literature review. *Front. Rehabil. Sci.* **2023**, *4*, 1122673. [CrossRef]
21. Ståhlberg, M.; Mahdi, A.; Johansson, M.; Fedorowski, A.; Olshansky, B. Cardiovascular dysautonomia in postacute sequelae of SARS-CoV-2 infection. *J. Cardiovasc. Electrophysiol.* 2023; *early view*. [CrossRef]
22. Gunning, W.T., 3rd; Kvale, H.; Kramer, P.M.; Karabin, B.L.; Grubb, B.P. Postural orthostatic tachycardia syndrome is associated with elevated g-protein coupled receptor autoantibodies. *J. Am. Heart Assoc.* **2019**, *8*, e013602. [CrossRef]
23. Blitshteyn, S. Autoimmune markers and autoimmune disorders in patients with postural tachycardia syndrome (POTS). *Lupus* **2015**, *24*, 1364–1369. [CrossRef]
24. Li, H.; Yu, X.; Liles, C.; Khan, M.; Vanderlinde-Wood, M.; Galloway, A.; Zillner, C.; Benbrook, A.; Reim, S.; Collier, D.; et al. Autoimmune basis for postural tachycardia syndrome. *J. Am. Heart Assoc.* **2014**, *3*, e000755. [CrossRef]
25. Dahan, S.; Tomljenovic, L.; Shoenfeld, Y. Postural orthostatic tachycardia syndrome (POTS)—A novel member of the autoimmune family. *Lupus* **2016**, *25*, 339–342. [CrossRef]
26. Vernino, S.; Stiles, L.E. Autoimmunity in postural orthostatic tachycardia syndrome: Current understanding. *Auton. Neurosci.* **2018**, *215*, 78–82. [CrossRef] [PubMed]
27. Seeley, M.C.; Hooper, M.; Tan, J.; Wells, R.; Gallagher, C.; Lau, D.H. Plasma exchange improves cognitive function in long-COVID-related postural orthostatic tachycardia syndrome and autoimmune neurological dysfunction. *Am. J. Med.* **2023**, *133*, e157–e159. [CrossRef] [PubMed]
28. Shanks, L.; Jason, L.A.; Evans, M.; Brown, A. Cognitive impairments associated with CFS and POTS. *Front. Physiol.* **2013**, *4*, 113. [CrossRef] [PubMed]
29. Arnold, A.C.; Haman, K.; Garland, E.M.; Raj, V.; Dupont, W.D.; Biaggioni, I.; Robertson, D.; Raj, S.R. Cognitive dysfunction in postural tachycardia syndrome. *Clin. Sci.* **2015**, *128*, 39–45. [CrossRef] [PubMed]
30. Raj, V.; Opie, M.; Arnold, A.C. Cognitive and psychological issues in postural tachycardia syndrome. *Auton. Neurosci.* **2018**, *215*, 46–55. [CrossRef]
31. Hoad, A.; Spickett, G.; Elliott, J.; Newton, J. Postural orthostatic tachycardia syndrome is an under-recognized condition in chronic fatigue syndrome. *QJM* **2008**, *101*, 961–965. [CrossRef]
32. Lewis, I.; Pairman, J.; Spickett, G.; Newton, J.L. Clinical characteristics of a novel subgroup of chronic fatigue syndrome patients with postural orthostatic tachycardia syndrome. *J. Intern. Med.* **2013**, *273*, 501–510. [CrossRef]
33. Nijs, J.; Ickmans, K. Postural orthostatic tachycardia syndrome as a clinically important subgroup of chronic fatigue syndrome: Further evidence for central nervous system dysfunctioning. *J. Intern. Med.* **2013**, *273*, 498–500. [CrossRef]
34. Reynolds, G.K.; Lewis, D.P.; Richardson, A.M.; Lidbury, B.A. Comorbidity of postural orthostatic tachycardia syndrome and chronic fatigue syndrome in an Australian cohort. *J. Intern. Med.* **2014**, *275*, 409–417. [CrossRef] [PubMed]
35. Wise, S.; Ross, A.; Brown, A.; Evans, M.; Jason, L. An assessment of fatigue in patients with postural orthostatic tachycardia syndrome. *J. Health Psychol.* **2017**, *22*, 733–742. [CrossRef] [PubMed]
36. Boris, J.R.; Bernadzikowski, T. Therapy for fatigue and cognitive dysfunction in postural orthostatic tachycardia syndrome. *Cardiol. Young* **2018**, *28*, 1415–1420. [CrossRef] [PubMed]
37. Strassheim, V.; Welford, J.; Ballantine, R.; Newton, J.L. Managing fatigue in postural tachycardia syndrome (PoTS): The newcastle approach. *Auton. Neurosci.* **2018**, *215*, 56–61. [CrossRef] [PubMed]
38. Astudillo, L.; Laure, A.; Fabry, V.; Pugnet, G.; Maury, P.; Labrunee, M.; Sailler, L.; Pavy-Le Traon, A. Postural tachycardia syndrome (PoTS): An up-to-date. *Rev. Med. Interne* **2018**, *39*, 627–634. [CrossRef]
39. Blitshteyn, S.; Fedorowski, A. The risks of POTS after COVID-19 vaccination and SARS-CoV-2 infection: More studies are needed. *Nat. Rev. Cardiol.* **2022**, *1*, 1119–1120. [CrossRef]
40. Blitshteyn, S. Human papillomavirus (HPV) vaccine safety concerning POTS, CRPS and related conditions. *Clin. Auton. Res.* **2020**, *30*, 181–182. [CrossRef]
41. Butts, B.N.; Fischer, P.R.; Mack, K.J. Human papillomavirus vaccine and postural orthostatic tachycardia syndrome: A review of current literature. *J. Child. Neurol.* **2017**, *32*, 956–965. [CrossRef]
42. Jefferson, T.; Jørgensen, L. Human papillomavirus vaccines, complex regional pain syndrome, postural orthostatic tachycardia syndrome, and autonomic dysfunction—A review of the regulatory evidence from the European medicines agency. *Indian J. Med. Ethics* **2017**, *2*, 30–37. [CrossRef] [PubMed]
43. Tomljenovic, L.; Colafrancesco, S.; Perricone, C.; Shoenfeld, Y. Postural orthostatic tachycardia with chronic fatigue after HPV vaccination as part of the "autoimmune/auto-inflammatory syndrome induced by adjuvants": Case report and literature review. *J. Investig. Med. High Impact Case Rep.* **2014**, *2*, 2324709614527812. [CrossRef]
44. Tv, P.; Tran, T.T.; Hao, H.T.; Hau, N.T.H.; Jain, N.; Reinis, A. Postural orthostatic tachycardia syndrome-like symptoms following COVID-19 vaccination: An overview of clinical literature. *Hum. Antibodies* **2023**, *31*, 9–17. [CrossRef]
45. Arana, J.; Mba-Jonas, A.; Jankosky, C.; Lewis, P.; Moro, P.L.; Shimabukuro, T.T.; Cano, M. Reports of postural orthostatic tachycardia syndrome after human papillomavirus vaccination in the vaccine adverse event reporting system. *J. Adolesc. Health* **2017**, *61*, 577–582. [CrossRef] [PubMed]

46. Barboi, A.; Gibbons, C.H.; Axelrod, F.; Benarroch, E.E.; Biaggioni, I.; Chapleau, M.W.; Chelimsky, G.; Chelimsky, T.; Cheshire, W.P.; Claydon, V.E.; et al. Human papillomavirus (HPV) vaccine and autonomic disorders: A position statement from the American autonomic society. *Clin. Auton. Res.* **2020**, *30*, 13–18. [CrossRef]
47. Gunning, W.T., 3rd; Stepkowski, S.M.; Kramer, P.M.; Karabin, B.L.; Grubb, B.P. Inflammatory biomarkers in postural orthostatic tachycardia syndrome with elevated G-protein-coupled receptor autoantibodies. *J. Clin. Med.* **2021**, *10*, 623. [CrossRef]
48. Mathias, C.J.; Low, D.A.; Iodice, V.; Owens, A.P.; Kirbis, M.; Grahame, R. Postural tachycardia syndrome—Current experience and concepts. *Nat. Rev. Neurol.* **2011**, *8*, 22–34. [CrossRef] [PubMed]
49. Davis, H.E.; Assaf, G.S.; McCorkell, L.; Wei, H.; Low, R.J.; Re'em, Y.; Redfield, S.; Austin, J.P.; Akrami, A. Characterizing long COVID in an international cohort: 7 months of symptoms and their impact. *eClinicalMedicine* **2021**, *38*, 101019. [CrossRef]
50. Mahdi, A.; Lodin, K.; Reistam, U.; Fedorowski, A.; Nygren-Bonnier, M.; Runold, M.; Bruchfeld, J.; Desta, L.; Pernow, J.; Nickander, J.; et al. Microvasular dysfunction and reduced cardiac stress reactivity in postural orthostatic tachycardia associated with postacute COVID-19. *Circ. Arrhythmia Electrophysiol.* **2023**, *16*, 413–414. [CrossRef]
51. Johansson, M.; Stahlberg, M.; Runold, M.; Nygren-Bonnier, M.; Nilsson, J.; Olshansky, B.; Bruchfeld, J.; Fedorowski, A. Long-haul post-COVID-19 symptoms presenting as a variant of postural orthostatic tachycardia syndrome: The swedish experience. *JACC Case Rep.* **2021**, *3*, 573–580. [CrossRef]
52. van Campen, C.L.M.C.; Rowe, P.C.; Visser, F.C. Orthostatic symptoms and reductions in cerebral blood flow in long-haul COVID-19 patients: Similarities with myalgic encephalomyelitis/chronic fatigue syndrome. *Medicina* **2021**, *58*, 28. [CrossRef]
53. Liu, L.D.; Duricka, D.L. Stellate ganglion block reduces symptoms of long COVID: A case series. *J. Neuroimmunol.* **2022**, *362*, 577784. [CrossRef]
54. Rigo, S.; Urechie, V.; Diedrich, A.; Okamoto, L.E.; Biaggioni, I.; Shibao, C.A. Impaired parasympathetic function in long-COVID postural orthostatic tachycardia syndrome—A case-control study. *Bioelectron. Med.* **2023**, *9*, 19. [CrossRef]
55. Seeley, M.C.; Gallagher, C.; Ong, E.; Langdon, A.; Chieng, J.; Bailey, D.; Page, A.; Lim, H.S.; Lau, D.H. High incidence of autonomic dysfunction and postural orthostatic tachycardia syndrome in patients with long COVID: Implications for management and health care planning. *Am. J. Med.* 2023; *in press*. [CrossRef]
56. Amekran, Y.; Damoun, N.; El Hangouche, A.J. Postural orthostatic tachycardia syndrome and post-acute COVID-19. *Glob. Cardiol. Sci. Pract.* **2022**, *2022*, e202213. [CrossRef] [PubMed]
57. Chadda, K.R.; Blakey, E.E.; Huang, C.L.; Jeevaratnam, K. Long COVID-19 and postural orthostatic tachycardia syndrome—Is dysautonomia to be blamed? *Front. Cardiovasc. Med.* **2022**, *9*, 860198. [CrossRef] [PubMed]
58. Ormiston, C.K.; Swiatkiewicz, I.; Taub, P.R. Postural orthostatic tachycardia syndrome as a sequela of COVID-19. *Heart Rhythm* **2022**, *19*, 1880–1889. [CrossRef]
59. Diekman, S.; Chung, T. Post-acute sequelae of SARS-CoV-2 syndrome presenting as postural orthostatic tachycardia syndrome. *Clin. Exp. Emerg. Med.* **2023**, *10*, 18–25. [CrossRef]
60. Gómez-Moyano, E.; Rodríguez-Capitán, J.; Gaitán Román, D.; Reyes Bueno, J.A.; Villalobos Sánchez, A.; Espildora Hernández, F.; González Angulo, G.E.; Molina Mora, M.J.; Thurnhofer-Hemsi, K.; Molina-Ramos, A.I.; et al. Postural orthostatic tachycardia syndrome and other related dysautonomic disorders after SARS-CoV-2 infection and after COVID-19 messenger RNA vaccination. *Front. Neurol.* **2023**, *14*, 1221518. [CrossRef]
61. Jamal, S.M.; Landers, D.B.; Hollenberg, S.M.; Turi, Z.G.; Glotzer, T.V.; Tancredi, J.; Parrillo, J.E. Prospective evaluation of autonomic dysfunction in post-acute sequela of COVID-19. *J. Am. Coll. Cardiol.* **2022**, *79*, 2325–2330. [CrossRef] [PubMed]
62. Minhas, R.; Bharadwaj, A.S. COVID-19-induced postural orthostatic tachycardia syndrome and dysautonomia. *Cureus* **2023**, *15*, e40235. [CrossRef]
63. Novak, P.; Giannetti, M.P.; Weller, E.; Hamilton, M.J.; Mukerji, S.S.; Alabsi, H.S.; Systrom, D.; Marciano, S.P.; Felsenstein, D.; Mullally, W.J.; et al. Network autonomic analysis of post-acute sequelae of COVID-19 and postural tachycardia syndrome. *Neurol. Sci.* **2022**, *43*, 6627–6638. [CrossRef] [PubMed]
64. Sherif, Z.A.; Gomez, C.R.; Connors, T.J.; Henrich, T.J.; Reeves, W.B.; Force, R.M.P.T. Pathogenic mechanisms of post-acute sequelae of SARS-CoV-2 infection (PASC). *eLife* **2023**, *12*, e86002. [CrossRef] [PubMed]
65. Novak, P.; Mukerji, S.S.; Alabsi, H.S.; Systrom, D.; Marciano, S.P.; Felsenstein, D.; Mullally, W.J.; Pilgrim, D.M. Multisystem involvement in post-acute sequelae of coronavirus disease 19. *Ann. Neurol.* **2022**, *91*, 367–379. [CrossRef] [PubMed]
66. Takao, M.; Ohira, M. Neurological post-acute sequelae of SARS-CoV-2 infection. *Psychiatry Clin. Neurosci.* **2023**, *77*, 72–83. [CrossRef]
67. Fedorowski, A.; Sutton, R. Autonomic dysfunction and postural orthostatic tachycardia syndrome in post-acute COVID-19 syndrome. *Nat. Rev. Cardiol.* **2023**, *20*, 281–282. [CrossRef] [PubMed]
68. Davis, H.E.; McCorkell, L.; Vogel, J.M.; Topol, E.J. Long COVID: Major findings, mechanisms and recommendations. *Nat. Rev. Microbiol.* **2023**, *21*, 133–146. [CrossRef]
69. Mueller, B.R.; Robinson-Papp, J. Postural orthostatic tachycardia syndrome and migraine: A narrative review. *Headache* **2022**, *62*, 792–800. [CrossRef]
70. Wekerle, H. Epstein-barr virus sparks brain autoimmunity in multiple sclerosis. *Nature* **2022**, *603*, 230–232. [CrossRef]
71. Kurtzke, J.F.; Heltberg, A. Multiple sclerosis in the faroe islands: An epitome. *J. Clin. Epidemiol.* **2001**, *54*, 1–22. [CrossRef]
72. Kurtzke, J.F. Epidemiology in multiple sclerosis: A pilgrim's progress. *Brain* **2013**, *136*, 2904–2917. [CrossRef] [PubMed]

73. Poser, C.M.; Hibberd, P.L. Analysis of the 'epidemic' of multiple sclerosis in the faroe islands. II. Biostatistical aspects. *Neuroepidemiology* **1988**, *7*, 181–189. [CrossRef]
74. Poser, C.M.; Hibberd, P.L.; Benedikz, J.; Gudmundsson, G. Analysis of the 'epidemic' of multiple sclerosis in the faroe islands. I. Clinical and epidemiological aspects. *Neuroepidemiology* **1988**, *7*, 168–180. [CrossRef] [PubMed]
75. Adamec, I.; Lovric, M.; Zaper, D.; Barusic, A.K.; Bach, I.; Junakovic, A.; Mismas, A.; Habek, M. Postural orthostatic tachycardia syndrome associated with multiple sclerosis. *Auton. Neurosci.* **2013**, *173*, 65–68. [CrossRef]
76. Proal, A.D.; VanElzakker, M.B. Long COVID or post-acute sequelae of COVID-19 (PASC): An overview of biological factors that may contribute to persistent symptoms. *Front. Microbiol.* **2021**, *12*, 698169. [CrossRef]
77. Altmann, D.M.; Whettlock, E.M.; Liu, S.; Arachchillage, D.J.; Boyton, R.J. The immunology of long COVID. *Nat. Rev. Immunol.* **2023**, *23*, 618–634. [CrossRef]
78. Komaroff, A.L.; Lipkin, W.I. ME/CFS and long COVID share similar symptoms and biological abnormalities: Road map to the literature. *Front. Med.* **2023**, *10*, 1187163. [CrossRef]
79. Ryabkova, V.A.; Gavrilova, N.Y.; Fedotkina, T.V.; Churilov, L.P.; Shoenfeld, Y. Myalgic encephalomyelitis/chronic fatigue syndrome and post-COVID syndrome: A common neuroimmune ground. *Diagnostics* **2023**, *13*, 66. [CrossRef]
80. van Campen, C.L.M.C.; Visser, F.C. The abnormal cardiac index and stroke volume index changes during a normal tilt table test in ME/CFS patients compared to healthy volunteers, are not related to deconditioning. *J. Thromb. Circ.* **2018**, *2018*, 108.
81. van Campen, C.; Rowe, P.C.; Visser, F.C. Blood volume status in ME/CFS correlates with the presence or absence of orthostatic symptoms: Preliminary results. *Front. Pediatr.* **2018**, *6*, 352. [CrossRef] [PubMed]
82. Bateman, L.; Bested, A.C.; Bonilla, H.F.; Chheda, B.V.; Chu, L.; Curtin, J.M.; Dempsey, T.T.; Dimmock, M.E.; Dowell, T.G.; Felsenstein, D.; et al. Myalgic encephalomyelitis/chronic fatigue syndrome: Essentials of diagnosis and management. *Mayo Clin. Proc.* **2021**, *96*, 2861–2878. [CrossRef] [PubMed]
83. van Campen, C.L.M.C.; Rowe, P.C.; Verheugt, F.W.A.; Visser, F.C. Influence of end-tidal CO_2 on cerebral blood flow during orthostatic stress in controls and adults with myalgic encephalomyelitis/chronic fatigue syndrome. *Physiol. Rep.* **2023**, *11*, e15639. [CrossRef] [PubMed]
84. van Campen, C.; Verheugt, F.W.A.; Rowe, P.C.; Visser, F.C. Orthostatic chronotropic incompetence in patients with myalgic encephalomyelitis/chronic fatigue syndrome (ME/CFS). *IBRO Neurosci. Rep.* **2023**, *15*, 1–10. [CrossRef] [PubMed]
85. Gunning, W.T.; Kramer, P.M.; Cichocki, J.A.; Karabin, B.L.; Khuder, S.A.; Grubb, B.P. Platelet storage pool deficiency and elevated inflammatory biomarkers are prevalent in postural orthostatic tachycardia syndrome. *Cells* **2022**, *11*, 774. [CrossRef] [PubMed]
86. Staff, A.C.; Redman, C.W.; Williams, D.; Leeson, P.; Moe, K.; Thilaganathan, B.; Magnus, P.; Steegers, E.A.; Tsigas, E.Z.; Ness, R.B.; et al. Pregnancy and long-term maternal cardiovascular health: Progress through harmonization of research cohorts and biobanks. *Hypertension* **2016**, *67*, 251–260. [CrossRef]
87. Thilaganathan, B.; Kalafat, E. Cardiovascular system in preeclampsia and beyond. *Hypertension* **2019**, *73*, 522–531. [CrossRef]
88. Morgan, K.; Chojenta, C.; Tavener, M.; Smith, A.; Loxton, D. Postural orthostatic tachycardia syndrome during pregnancy: A systematic review of the literature. *Auton. Neurosci.* **2018**, *215*, 106–118. [CrossRef]
89. Morgan, K.; Smith, A.; Blitshteyn, S. POTS and pregnancy: A review of literature and recommendations for evaluation and treatment. *Int. J. Womens Health* **2022**, *14*, 1831–1847. [CrossRef]
90. Raj, S.R.; Guzman, J.C.; Harvey, P.; Richer, L.; Schondorf, R.; Seifer, C.; Thibodeau-Jarry, N.; Sheldon, R.S. Canadian cardiovascular society position statement on postural orthostatic tachycardia syndrome (POTS) and related disorders of chronic orthostatic intolerance. *Can. J. Cardiol.* **2020**, *36*, 357–372. [CrossRef]
91. Zadourian, A.; Doherty, T.A.; Swiatkiewicz, I.; Taub, P.R. Postural orthostatic tachycardia syndrome: Prevalence, pathophysiology, and management. *Drugs* **2018**, *78*, 983–994. [CrossRef]
92. Gordan, R.; Gwathmey, J.K.; Xie, L.H. Autonomic and endocrine control of cardiovascular function. *World J. Cardiol.* **2015**, *7*, 204–214. [CrossRef] [PubMed]
93. Armstrong, R.; Wheen, P.; Brandon, L.; Maree, A.; Kenny, R.A. Heart rate: Control mechanisms, pathophysiology and assessment of the neurocardiac system in health and disease. *QJM* **2022**, *115*, 806–812. [CrossRef] [PubMed]
94. Simon, P.M.; Taha, B.H.; Dempsey, J.A.; Skatrud, J.B.; Iber, C. Role of vagal feedback from the lung in hypoxic-induced tachycardia in humans. *J. Appl. Physiol.* **1995**, *78*, 1522–1530. [CrossRef] [PubMed]
95. Faulhaber, M.; Gatterer, H.; Haider, T.; Linser, T.; Netzer, N.; Burtscher, M. Heart rate and blood pressure responses during hypoxic cycles of a 3-week intermittent hypoxia breathing program in patients at risk for or with mild COPD. *Int. J. Chronic Obstr. Pulm. Dis.* **2015**, *10*, 339–345. [CrossRef] [PubMed]
96. Siebenmann, C.; Lundby, C. Regulation of cardiac output in hypoxia. *Scand. J. Med. Sci. Sports* **2015**, *25* (Suppl. S4), 53–59. [CrossRef] [PubMed]
97. Paleczny, B.; Seredynski, R.; Tubek, S.; Adamiec, D.; Ponikowski, P.; Ponikowska, B. Hypoxic tachycardia is not a result of increased respiratory activity in healthy subjects. *Exp. Physiol.* **2019**, *104*, 476–489. [CrossRef] [PubMed]
98. Joyce, W.; Wang, T. Regulation of heart rate in vertebrates during hypoxia: A comparative overview. *Acta Physiol.* **2022**, *234*, e13779. [CrossRef]
99. Kell, D.B.; Pretorius, E. No effects without causes. The iron dysregulation and dormant microbes hypothesis for chronic, inflammatory diseases. *Biol. Rev.* **2018**, *93*, 1518–1557. [CrossRef]

100. Deer, R.R.; Rock, M.A.; Vasilevsky, N.; Carmody, L.; Rando, H.; Anzalone, A.J.; Basson, M.D.; Bennett, T.D.; Bergquist, T.; Boudreau, E.A.; et al. Characterizing long COVID: Deep phenotype of a complex condition. *eBioMedicine* **2021**, *74*, 103722. [CrossRef]
101. Reese, J.T.; Blau, H.; Casiraghi, E.; Bergquist, T.; Loomba, J.J.; Callahan, T.J.; Laraway, B.; Antonescu, C.; Coleman, B.; Gargano, M.; et al. Generalisable long COVID subtypes: Findings from the NIH N3C and recover programmes. *eBioMedicine* **2023**, *87*, 104413. [CrossRef] [PubMed]
102. Yong, S.J.; Liu, S. Proposed subtypes of post-COVID-19 syndrome (or long-COVID) and their respective potential therapies. *Rev. Med. Virol.* **2022**, *32*, e2315. [CrossRef]
103. Low, P.A.; Schondorf, R.; Rummans, T.A. Why do patients have orthostatic symptoms in POTS? *Clin. Auton. Res.* **2001**, *11*, 223–224. [CrossRef] [PubMed]
104. Plash, W.B.; Diedrich, A.; Biaggioni, I.; Garland, E.M.; Paranjape, S.Y.; Black, B.K.; Dupont, W.D.; Raj, S.R. Diagnosing postural tachycardia syndrome: Comparison of tilt testing compared with standing haemodynamics. *Clin. Sci.* **2013**, *124*, 109–114. [CrossRef] [PubMed]
105. Novak, P. Cerebral blood flow, heart rate, and blood pressure patterns during the tilt test in common orthostatic syndromes. *Neurosci. J.* **2016**, *2016*, 6127340. [CrossRef] [PubMed]
106. Aponte-Becerra, L.; Novak, P. Tilt test: A review. *J. Clin. Neurophysiol.* **2021**, *38*, 279–286. [CrossRef]
107. Cheshire, W.P., Jr.; Goldstein, D.S. Autonomic uprising: The tilt table test in autonomic medicine. *Clin. Auton. Res.* **2019**, *29*, 215–230. [CrossRef] [PubMed]
108. Stewart, J.M.; Visintainer, P.; Medow, M.S. Standing tests lack reliability to diagnose all adolescents who have postural tachycardia syndrome. *Clin. Auton. Res.* **2023**, *33*, 899–901. [CrossRef] [PubMed]
109. Petersen, M.E.V.; Williams, T.R.; Gordon, C.; Chamberlain-Webber, R.; Sutton, R. The normal response to prolonged passive head up tilt testing. *Heart* **2000**, *84*, 509–514. [CrossRef]
110. Purkayastha, S.; Sorond, F. Transcranial doppler ultrasound: Technique and application. *Semin. Neurol.* **2012**, *32*, 411–420. [CrossRef]
111. Thijs, R.D.; Brignole, M.; Falup-Pecurariu, C.; Fanciulli, A.; Freeman, R.; Guaraldi, P.; Jordan, J.; Habek, M.; Hilz, M.; Traon, A.P.; et al. Recommendations for tilt table testing and other provocative cardiovascular autonomic tests in conditions that may cause transient loss of consciousness: Consensus statement of the European federation of autonomic societies (EFAS) endorsed by the American autonomic society (AAS) and the European academy of neurology (EAN). *Clin. Auton. Res.* **2021**, *31*, 369–384. [CrossRef]
112. Raj, S.R.; Fedorowski, A.; Sheldon, R.S. Diagnosis and management of postural orthostatic tachycardia syndrome. *CMAJ* **2022**, *194*, E378–E385. [CrossRef] [PubMed]
113. Spahic, J.M.; Hamrefors, V.; Johansson, M.; Ricci, F.; Melander, O.; Sutton, R.; Fedorowski, A. Malmö POTS symptom score: Assessing symptom burden in postural orthostatic tachycardia syndrome. *J. Intern. Med.* **2023**, *293*, 91–99. [CrossRef] [PubMed]
114. Benarroch, E.E. Postural tachycardia syndrome: A heterogeneous and multifactorial disorder. *Mayo Clin. Proc.* **2012**, *87*, 1214–1225. [CrossRef]
115. Petelin Gadze, Z.; Bujan Kovac, A.; Adamec, I.; Milekic, N.; Sulentic, V. Vagal nerve stimulation is beneficial in postural orthostatic tachycardia syndrome and epilepsy. *Seizure* **2018**, *57*, 11–13. [CrossRef]
116. Diedrich, A.; Urechie, V.; Shiffer, D.; Rigo, S.; Minonzio, M.; Cairo, B.; Smith, E.C.; Okamoto, L.E.; Barbic, F.; Bisoglio, A.; et al. Transdermal auricular vagus stimulation for the treatment of postural tachycardia syndrome. *Auton. Neurosci.* **2021**, *236*, 102886. [CrossRef]
117. Benarroch, E.E. "Dysautonomia": A plea for precision. *Clin. Auton. Res.* **2021**, *31*, 27–29. [CrossRef]
118. Palma, J.A.; Norcliffe-Kaufmann, L.; Fuente-Mora, C.; Percival, L.; Mendoza-Santiesteban, C.; Kaufmann, H. Current treatments in familial dysautonomia. *Expert. Opin. Pharmacother.* **2014**, *15*, 2653–2671. [CrossRef] [PubMed]
119. Murga, I.; Aranburu, L.; Gargiulo, P.A.; Gomez Esteban, J.C.; Lafuente, J.V. Clinical heterogeneity in ME/CFS. A way to understand long-COVID19 fatigue. *Front. Psychiatry* **2021**, *12*, 735784. [CrossRef]
120. Carmona-Torre, F.; Minguez-Olaondo, A.; Lopez-Bravo, A.; Tijero, B.; Grozeva, V.; Walcker, M.; Azkune-Galparsoro, H.; Lopez de Munain, A.; Alcaide, A.B.; Quiroga, J.; et al. Dysautonomia in COVID-19 patients: A narrative review on clinical course, diagnostic and therapeutic strategies. *Front. Neurol.* **2022**, *13*, 886609. [CrossRef]
121. Bisaccia, G.; Ricci, F.; Recce, V.; Serio, A.; Iannetti, G.; Chahal, A.A.; Stahlberg, M.; Khanji, M.Y.; Fedorowski, A.; Gallina, S. Post-acute sequelae of COVID-19 and cardiovascular autonomic dysfunction: What do we know? *J. Cardiovasc. Dev. Dis.* **2021**, *8*, 156. [CrossRef]
122. Chung, T.H.; Azar, A. Autonomic nerve involvement in post-acute sequelae of SARS-CoV-2 syndrome (PASC). *J. Clin. Med.* **2022**, *12*, 73. [CrossRef] [PubMed]
123. Marques, K.C.; Quaresma, J.A.S.; Falcao, L.F.M. Cardiovascular autonomic dysfunction in "long COVID": Pathophysiology, heart rate variability, and inflammatory markers. *Front. Cardiovasc. Med.* **2023**, *10*, 1256512. [CrossRef]
124. Finsterer, J. Small fiber neuropathy underlying dysautonomia in COVID-19 and in post-SARS-CoV-2 vaccination and long-COVID syndromes. *Muscle Nerve* **2022**, *65*, E31–E32. [CrossRef] [PubMed]

125. Aktürk, T.; Turan, Y.; Tanik, N.; Karadağ, M.E.; Sacmaci, H.; Inan, L.E. Vitamin D, vitamin D binding protein, vitamin D receptor levels and cardiac dysautonomia in patients with multiple sclerosis: A cross-sectional study. *Arq. Neuro-Psiquiatr.* **2019**, *77*, 848–854. [CrossRef]
126. Foschi, M.; Giannini, G.; Merli, E.; Mancinelli, L.; Zenesini, C.; Viti, B.; Guaraldi, P.; Cortelli, P.; Lugaresi, A. Frequency and characteristics of dysautonomic symptoms in multiple sclerosis: A cross-sectional double-center study with the validated Italian version of the composite autonomic symptom score-31. *Neurol. Sci.* **2021**, *42*, 1395–1403. [CrossRef]
127. Stewart, J.M. Autonomic nervous system dysfunction in adolescents with postural orthostatic tachycardia syndrome and chronic fatigue syndrome is characterized by attenuated vagal baroreflex and potentiated sympathetic vasomotion. *Pediatr. Res.* **2000**, *48*, 218–226. [CrossRef] [PubMed]
128. Naschitz, J.E.; Yeshurun, D.; Rosner, I. Dysautonomia in chronic fatigue syndrome: Facts, hypotheses, implications. *Med. Hypotheses* **2004**, *62*, 203–206. [CrossRef]
129. Bested, A.C.; Marshall, L.M. Review of myalgic encephalomyelitis/chronic fatigue syndrome: An evidence-based approach to diagnosis and management by clinicians. *Rev. Environ. Health* **2015**, *30*, 223–249. [CrossRef] [PubMed]
130. Nelson, M.J.; Bahl, J.S.; Buckley, J.D.; Thomson, R.L.; Davison, K. Evidence of altered cardiac autonomic regulation in myalgic encephalomyelitis/chronic fatigue syndrome: A systematic review and meta-analysis. *Medicine* **2019**, *98*, e17600. [CrossRef]
131. Murga Gandasegui, I.; Aranburu Laka, L.; Gargiulo, P.Á.; Gómez-Esteban, J.C.; Lafuente Sánchez, J.V. Myalgic encephalomyelitis/chronic fatigue syndrome: A neurological entity? *Medicina* **2021**, *57*, 1030. [CrossRef]
132. Nelson, M.J.; Buckley, J.D.; Thomson, R.L.; Bellenger, C.R.; Davison, K. Markers of cardiac autonomic function during consecutive day peak exercise tests in people with myalgic encephalomyelitis/chronic fatigue syndrome. *Front. Physiol.* **2021**, *12*, 771899. [CrossRef]
133. Renz-Polster, H.; Tremblay, M.E.; Bienzle, D.; Fischer, J.E. The pathobiology of myalgic encephalomyelitis/chronic fatigue syndrome: The case for neuroglial failure. *Front. Cell. Neurosci.* **2022**, *16*, 888232. [CrossRef] [PubMed]
134. Pavy-Le Traon, A.; Amarenco, G.; Duerr, S.; Kaufmann, H.; Lahrmann, H.; Shaftman, S.R.; Tison, F.; Wenning, G.K.; Goetz, C.G.; Poewe, W.; et al. The movement disorders task force review of dysautonomia rating scales in Parkinson's disease with regard to symptoms of orthostatic hypotension. *Mov. Disord.* **2011**, *26*, 1985–1992. [CrossRef]
135. Krupp, L.B. Fatigue in multiple sclerosis: Definition, pathophysiology and treatment. *CNS Drugs* **2003**, *17*, 225–234. [CrossRef] [PubMed]
136. Shen, J.; Barbera, J.; Shapiro, C.M. Distinguishing sleepiness and fatigue: Focus on definition and measurement. *Sleep. Med. Rev.* **2006**, *10*, 63–76. [CrossRef]
137. Choi-Kwon, S.; Kim, J.S. Poststroke fatigue: An emerging, critical issue in stroke medicine. *Int. J. Stroke* **2011**, *6*, 328–336. [CrossRef]
138. Kluger, B.M.; Herlofson, K.; Chou, K.L.; Lou, J.S.; Goetz, C.G.; Lang, A.E.; Weintraub, D.; Friedman, J. Parkinson's disease-related fatigue: A case definition and recommendations for clinical research. *Mov. Disord.* **2016**, *31*, 625–631. [CrossRef] [PubMed]
139. Poulsen, M.B.; Skovbølling, S.L.; Kruuse, C.; Overgaard, K.; Rasmussen, R.S. How to identify fatigue in stroke patients: An investigation of the post-stroke fatigue case definition validity. *Top. Stroke Rehabil.* **2020**, *27*, 369–376. [CrossRef]
140. Moore, Y.; Serafimova, T.; Anderson, N.; King, H.; Richards, A.; Brigden, A.; Sinai, P.; Higgins, J.; Ascough, C.; Clery, P.; et al. Recovery from chronic fatigue syndrome: A systematic review-heterogeneity of definition limits study comparison. *Arch. Dis. Child.* **2021**, *106*, 1087–1094. [CrossRef]
141. Adibi, I.; Sanayei, M.; Tabibian, F.; Ramezani, N.; Pourmohammadi, A.; Azimzadeh, K. Multiple sclerosis-related fatigue lacks a unified definition: A narrative review. *J. Res. Med. Sci.* **2022**, *27*, 24. [CrossRef]
142. Levesque, A.; Caru, M.; Duval, M.; Laverdiere, C.; Marjerrison, S.; Sultan, S. Cancer-related fatigue: Scoping review to synthesize a definition for childhood cancer survivors. *Support. Care Cancer* **2023**, *31*, 231. [CrossRef] [PubMed]
143. Pretorius, E.; Bronkhorst, P.; Briedenhann, S.; Smit, E.; Franz, R.C. Comparisons of the fibrin networks during pregnancy, nonpregnancy and pregnancy during dysfibrinogenaemia using the scanning electron microscope. *Blood Coagul. Fibrinolysis* **2009**, *20*, 12–16. [CrossRef] [PubMed]
144. Pretorius, E.; Oberholzer, H.M.; van der Spuy, W.J.; Meiring, J.H. The changed ultrastructure of fibrin networks during use of oral contraception and hormone replacement. *J. Thromb. Thrombolysis* **2010**, *30*, 502–506. [CrossRef] [PubMed]
145. Pretorius, E.; Steyn, H.; Engelbrecht, M.; Swanepoel, A.C.; Oberholzer, H.M. Differences in fibrin fiber diameters in healthy individuals and thromboembolic ischemic stroke patients. *Blood Coagul. Fibrinolysis* **2011**, *22*, 696–700. [CrossRef] [PubMed]
146. Pretorius, E.; Vermeulen, N.; Bester, J.; Lipinski, B.; Kell, D.B. A novel method for assessing the role of iron and its functional chelation in fibrin fibril formation: The use of scanning electron microscopy. *Toxicol. Mech. Methods* **2013**, *23*, 352–359. [CrossRef] [PubMed]
147. Pretorius, E.; Mbotwe, S.; Bester, J.; Robinson, C.J.; Kell, D.B. Acute induction of anomalous and amyloidogenic blood clotting by molecular amplification of highly substoichiometric levels of bacterial lipopolysaccharide. *J. R. Soc. Interface* **2016**, *123*, 20160539. [CrossRef]
148. Kell, D.B.; Pretorius, E. Proteins behaving badly. Substoichiometric molecular control and amplification of the initiation and nature of amyloid fibril formation: Lessons from and for blood clotting. *Prog. Biophys. Mol. Biol.* **2017**, *123*, 16–41. [CrossRef]
149. Pretorius, E.; Mbotwe, S.; Kell, D.B. Lipopolysaccharide-binding protein (LBP) reverses the amyloid state of fibrin seen in plasma of type 2 diabetics with cardiovascular comorbidities. *Sci. Rep.* **2017**, *7*, 9680. [CrossRef]

150. Pretorius, E.; Page, M.J.; Hendricks, L.; Nkosi, N.B.; Benson, S.R.; Kell, D.B. Both lipopolysaccharide and lipoteichoic acids potently induce anomalous fibrin amyloid formation: Assessment with novel Amytracker™ stains. *J. R. Soc. Interface* **2018**, *15*, 20170941. [CrossRef]
151. Pretorius, E.; Page, M.J.; Mbotwe, S.; Kell, D.B. Lipopolysaccharide-binding protein (LBP) can reverse the amyloid state of fibrin seen or induced in Parkinson's disease. *PLoS ONE* **2018**, *13*, e0192121. [CrossRef]
152. Pretorius, E.; Bester, J.; Page, M.J.; Kell, D.B. The potential of LPS-binding protein to reverse amyloid formation in plasma fibrin of individuals with Alzheimer-type dementia. *Front. Aging Neurosci* **2018**, *10*, 257. [CrossRef]
153. de Waal, G.M.; Engelbrecht, L.; Davis, T.; de Villiers, W.J.S.; Kell, D.B.; Pretorius, E. Correlative light-electron microscopy detects lipopolysaccharide and its association with fibrin fibres in parkinson's disease, Alzheimer's disease and type 2 diabetes mellitus. *Sci. Rep.* **2018**, *8*, 16798. [CrossRef] [PubMed]
154. Laubscher, G.J.; Lourens, P.J.; Venter, C.; Kell, D.B.; Pretorius, E. TEG®, microclot and platelet mapping for guiding early management of severe COVID-19 coagulopathy. *J. Clin. Med.* **2021**, *10*, 5381. [CrossRef]
155. Pretorius, E.; Kell, D.B. A perspective on how microscopy imaging of fibrinaloid microclots and platelet pathology may be applied in clinical investigations. In *Seminars in Thrombosis and Hemostasis*; Thieme Medical Publishers, Inc.: New York, NY, USA, 2023. [CrossRef]
156. Pretorius, E.; Bester, J.; Vermeulen, N.; Alummoottil, S.; Soma, P.; Buys, A.V.; Kell, D.B. Poorly controlled type 2 diabetes is accompanied by significant morphological and ultrastructural changes in both erythrocytes and in thrombin-generated fibrin: Implications for diagnostics. *Cardiovasc. Diabetol.* **2015**, *13*, 30. [CrossRef] [PubMed]
157. Pretorius, E.; Page, M.J.; Engelbrecht, L.; Ellis, G.C.; Kell, D.B. Substantial fibrin amyloidogenesis in type 2 diabetes assessed using amyloid-selective fluorescent stains. *Cardiovasc. Diabetol.* **2017**, *16*, 141. [CrossRef]
158. Pretorius, E.; Venter, C.; Laubscher, G.J.; Lourens, P.J.; Steenkamp, J.; Kell, D.B. Prevalence of readily detected amyloid blood clots in 'unclotted' type 2 diabetes mellitus and COVID-19 plasma: A preliminary report. *Cardiovasc. Diabetol.* **2020**, *19*, 193. [CrossRef] [PubMed]
159. Kell, D.B.; Laubscher, G.J.; Pretorius, E. A central role for amyloid fibrin microclots in long COVID/PASC: Origins and therapeutic implications. *Biochem. J.* **2022**, *479*, 537–559. [CrossRef]
160. Kell, D.B.; Pretorius, E. The potential role of ischaemia-reperfusion injury in chronic, relapsing diseases such as rheumatoid arthritis, long COVID and ME/CFS: Evidence, mechanisms, and therapeutic implications. *Biochem. J.* **2022**, *479*, 1653–1708. [CrossRef]
161. Kell, D.B.; Pretorius, E. Are fibrinaloid microclots a cause of autoimmunity in long COVID and other post-infection diseases? *Biochem. J.* **2023**, *480*, 1217–1240. [CrossRef]
162. Kruger, A.; Vlok, M.; Turner, S.; Venter, C.; Laubscher, G.J.; Kell, D.B.; Pretorius, E. Proteomics of fibrin amyloid microclots in long COVID/post-acute sequelae of COVID-19 (PASC) shows many entrapped pro-inflammatory molecules that may also contribute to a failed fibrinolytic system. *Cardiovasc. Diabetol.* **2022**, *21*, 190. [CrossRef]
163. Pretorius, E.; Vlok, M.; Venter, C.; Bezuidenhout, J.A.; Laubscher, G.J.; Steenkamp, J.; Kell, D.B. Persistent clotting protein pathology in long COVID/post-acute sequelae of COVID-19 (PASC) is accompanied by increased levels of antiplasmin. *Cardiovasc. Diabetol.* **2021**, *20*, 172. [CrossRef]
164. Pretorius, E.; Venter, C.; Laubscher, G.J.; Kotze, M.J.; Oladejo, S.; Watson, L.R.; Rajaratnam, K.; Watson, B.W.; Kell, D.B. Prevalence of symptoms, comorbidities, fibrin amyloid microclots and platelet pathology in individuals with long COVID/post-acute sequelae of COVID-19 (PASC). *Cardiovasc. Diabetol.* **2022**, *21*, 148. [CrossRef]
165. Turner, S.; Khan, M.A.; Putrino, D.; Woodcock, A.; Kell, D.B.; Pretorius, E. Long COVID: Pathophysiological factors and abnormal coagulation. *Trends Endocrinol. Metab.* **2023**, *34*, 321–344. [CrossRef]
166. Turner, S.; Laubscher, G.J.; Khan, M.A.; Kell, D.B.; Pretorius, E. Accelerating discovery: A novel flow cytometric method for detecting fibrin(ogen) amyloid microclots using long COVID as a model. *Heliyon* **2023**, *9*, e19605. [CrossRef] [PubMed]
167. Grobbelaar, L.M.; Venter, C.; Vlok, M.; Ngoepe, M.; Laubscher, G.J.; Lourens, P.J.; Steenkamp, J.; Kell, D.B.; Pretorius, E. SARS-CoV-2 spike protein S1 induces fibrin(ogen) resistant to fibrinolysis: Implications for microclot formation in COVID-19. *Biosci. Rep.* **2021**, *41*, BSR20210611. [CrossRef] [PubMed]
168. Grobbelaar, L.M.; Kruger, A.; Venter, C.; Burger, E.M.; Laubscher, G.J.; Maponga, T.G.; Kotze, M.J.; Kwaan, H.C.; Miller, J.B.; Fulkerson, D.; et al. Relative hypercoagulopathy of the SARS-CoV-2 beta and delta variants when compared to the less severe omicron variants is related to TEG parameters, the extent of fibrin amyloid microclots, and the severity of clinical illness. *Semin. Thromb. Haemost.* **2022**, *48*, 858–868. [CrossRef]
169. Nunes, J.M.; Kruger, A.; Proal, A.; Kell, D.B.; Pretorius, E. The occurrence of hyperactivated platelets and fibrinaloid microclots in myalgic encephalomyelitis/chronic fatigue syndrome (ME/CFS). *Pharmaceuticals* **2022**, *15*, 931. [CrossRef] [PubMed]
170. Nunes, J.M.; Kell, D.B.; Pretorius, E. Cardiovascular and haematological pathology in myalgic encephalomyelitis/chronic fatigue syndrome (ME/CFS): A role for viruses. *Blood Rev.* **2023**, *60*, 101075. [CrossRef] [PubMed]
171. Kell, D.B.; Pretorius, E. The simultaneous occurrence of both hypercoagulability and hypofibrinolysis in blood and serum during systemic inflammation, and the roles of iron and fibrin(ogen). *Integr. Biol.* **2015**, *7*, 24–52. [CrossRef]
172. Oaklander, A.L.; Mills, A.J.; Kelley, M.; Toran, L.S.; Smith, B.; Dalakas, M.C.; Nath, A. Peripheral neuropathy evaluations of patients with prolonged long COVID. *Neurol.-Neuroimmunol. Neuroinflamm.* **2022**, *9*, e1146. [CrossRef]

173. Pretorius, E.; Venter, C.; Laubscher, G.J.; Lourens, P.J.; Steenkamp, J.; Kell, D.B. Prevalence of amyloid blood clots in COVID-19 plasma. *medRxiv* **2020**. [CrossRef]
174. Safavi, F.; Gustafson, L.; Walitt, B.; Lehky, T.; Dehbashi, S.; Wiebold, A.; Mina, Y.; Shin, S.; Pan, B.; Polydefkis, M.; et al. Neuropathic symptoms with SARS-CoV-2 vaccination. *medRxiv* **2022**. [CrossRef]
175. Theoharides, T.C. Could SARS-CoV-2 spike protein be responsible for long-COVID syndrome? *Mol. Neurobiol.* **2022**, *59*, 1850–1861. [CrossRef]
176. Theoharides, T.C.; Kempuraj, D. Role of SARS-CoV-2 spike-protein-induced activation of microglia and mast cells in the pathogenesis of neuro-COVID. *Cells* **2023**, *12*, 688. [CrossRef]
177. Roy, M.; Nath, A.K.; Pal, I.; Dey, S.G. Second sphere interactions in amyloidogenic diseases. *Chem. Rev.* **2022**, *122*, 12132–12206. [CrossRef]
178. Janson, J.; Ashley, R.H.; Harrison, D.; McIntyre, S.; Butler, P.C. The mechanism of islet amyloid polypeptide toxicity is membrane disruption by intermediate-sized toxic amyloid particles. *Diabetes* **1999**, *48*, 491–498. [CrossRef]
179. Engel, M.F.M.; Khemtémourian, L.; Kleijer, C.C.; Meeldijk, H.J.D.; Jacobs, J.; Verkleij, A.J.; de Kruijff, B.; Killian, J.A.; Hoppener, J.W.M. Membrane damage by human islet amyloid polypeptide through fibril growth at the membrane. *Proc. Natl. Acad. Sci. USA* **2008**, *105*, 6033–6038. [CrossRef]
180. Friedman, R.; Pellarin, R.; Caflisch, A. Amyloid aggregation on lipid bilayers and its impact on membrane permeability. *J. Mol. Biol.* **2009**, *387*, 407–415. [CrossRef]
181. Butterfield, S.M.; Lashuel, H.A. Amyloidogenic protein-membrane interactions: Mechanistic insight from model systems. *Angew. Chem. Int. Ed. Engl.* **2010**, *49*, 5628–5654. [CrossRef]
182. Sciacca, M.F.M.; Brender, J.R.; Lee, D.K.; Ramamoorthy, A. Phosphatidylethanolamine enhances amyloid fiber-dependent membrane fragmentation. *Biochemistry* **2012**, *51*, 7676–7684. [CrossRef]
183. Sciacca, M.F.M.; Kotler, S.A.; Brender, J.R.; Chen, J.; Lee, D.K.; Ramamoorthy, A. Two-step mechanism of membrane disruption by abeta through membrane fragmentation and pore formation. *Biophys. J.* **2012**, *103*, 702–710. [CrossRef]
184. Relini, A.; Marano, N.; Gliozzi, A. Misfolding of amyloidogenic proteins and their interactions with membranes. *Biomolecules* **2013**, *4*, 20–55. [CrossRef]
185. Sciacca, M.F.M.; Milardi, D.; Messina, G.M.L.; Marletta, G.; Brender, J.R.; Ramamoorthy, A.; La Rosa, C. Cations as switches of amyloid-mediated membrane disruption mechanisms: Calcium and IAPP. *Biophys. J.* **2013**, *104*, 173–184. [CrossRef]
186. Jang, H.; Arce, F.T.; Ramachandran, S.; Kagan, B.L.; Lal, R.; Nussinov, R. Disordered amyloidogenic peptides may insert into the membrane and assemble into common cyclic structural motifs. *Chem. Soc. Rev.* **2014**, *43*, 6750–6764. [CrossRef]
187. Ow, S.Y.; Dunstan, D.E. A brief overview of amyloids and Alzheimer's disease. *Protein Sci.* **2014**, *23*, 1315–1331. [CrossRef]
188. Relini, A.; Marano, N.; Gliozzi, A. Probing the interplay between amyloidogenic proteins and membranes using lipid monolayers and bilayers. *Adv. Colloid Interface Sci.* **2014**, *207*, 81–92. [CrossRef]
189. Caillon, L.; Hoffmann, A.R.; Botz, A.; Khemtemourian, L. Molecular structure, membrane interactions, and toxicity of the islet amyloid polypeptide in type 2 diabetes mellitus. *J. Diabetes Res.* **2016**, *2016*, 5639875. [CrossRef]
190. Bode, D.C.; Baker, M.D.; Viles, J.H. Ion channel formation by amyloid-beta42 oligomers but not amyloid-beta40 in cellular membranes. *J. Biol. Chem.* **2017**, *292*, 1404–1413. [CrossRef]
191. Malishev, R.; Shaham-Niv, S.; Nandi, S.; Kolusheva, S.; Gazit, E.; Jelinek, R. Bacoside-A, an indian traditional-medicine substance, inhibits beta-amyloid cytotoxicity, fibrillation, and membrane interactions. *ACS Chem. Neurosci.* **2017**, *8*, 884–891. [CrossRef]
192. Cheng, B.; Li, Y.; Ma, L.; Wang, Z.; Petersen, R.B.; Zheng, L.; Chen, Y.; Huang, K. Interaction between amyloidogenic proteins and biomembranes in protein misfolding diseases: Mechanisms, contributors, and therapy. *Biochim. Biophys. Acta Biomembr.* **2018**, *1860*, 1876–1888. [CrossRef]
193. Rawat, A.; Langen, R.; Varkey, J. Membranes as modulators of amyloid protein misfolding and target of toxicity. *Biochim. Biophys. Acta Biomembr.* **2018**, *1860*, 1863–1875. [CrossRef]
194. Sciacca, M.F.M.; Chillemi, R.; Sciuto, S.; Greco, V.; Messineo, C.; Kotler, S.A.; Lee, D.K.; Brender, J.R.; Ramamoorthy, A.; La Rosa, C.; et al. A blend of two resveratrol derivatives abolishes hIAPP amyloid growth and membrane damage. *Biochim. Biophys. Acta Biomembr.* **2018**, *1860*, 1793–1802. [CrossRef]
195. Younan, N.D.; Chen, K.F.; Rose, R.S.; Crowther, D.C.; Viles, J.H. Prion protein stabilizes amyloid-beta (abeta) oligomers and enhances abeta neurotoxicity in a *Drosophila* model of Alzheimer's disease. *J. Biol. Chem.* **2018**, *293*, 13090–13099. [CrossRef]
196. Bode, D.C.; Freeley, M.; Nield, J.; Palma, M.; Viles, J.H. Amyloid-beta oligomers have a profound detergent-like effect on lipid membrane bilayers, imaged by atomic force and electron microscopy. *J. Biol. Chem.* **2019**, *294*, 7566–7572. [CrossRef]
197. Malishev, R.; Kolusheva, S.; Jelinek, R. Vesicle-based assays to study membrane interactions of amyloid peptides. *Methods Mol. Biol.* **2019**, *1873*, 39–51. [CrossRef]
198. Huang, Y.R.; Liu, R.T. The toxicity and polymorphism of beta-amyloid oligomers. *Int. J. Mol. Sci.* **2020**, *21*, 4477. [CrossRef]
199. Sciacca, M.F.M.; La Rosa, C.; Milardi, D. Amyloid-mediated mechanisms of membrane disruption. *Biophysica* **2021**, *1*, 137–156. [CrossRef]
200. Tempra, C.; Scollo, F.; Pannuzzo, M.; Lolicato, F.; La Rosa, C. A unifying framework for amyloid-mediated membrane damage: The lipid-chaperone hypothesis. *Biochim. Biophys. Acta Proteins Proteom.* **2022**, *1870*, 140767. [CrossRef]
201. Ma, L.; Li, X.; Petersen, R.B.; Peng, A.; Huang, K. Probing the interactions between amyloidogenic proteins and bio-membranes. *Biophys. Chem.* **2023**, *296*, 106984. [CrossRef]

202. Viles, J.H. Imaging amyloid-beta membrane interactions: Ion-channel pores and lipid-bilayer permeability in Alzheimer's disease. *Angew. Chem. Int. Ed. Engl.* **2023**, *62*, e202215785. [CrossRef]
203. Busciglio, J.; Lorenzo, A.; Yankner, B.A. Methodological variables in the assessment of beta amyloid neurotoxicity. *Neurobiol. Aging* **1992**, *13*, 609–612. [CrossRef]
204. Ono, K.; Condron, M.M.; Teplow, D.B. Structure-neurotoxicity relationships of amyloid beta-protein oligomers. *Proc. Natl. Acad. Sci. USA* **2009**, *106*, 14745–14750. [CrossRef]
205. Broersen, K.; Rousseau, F.; Schymkowitz, J. The culprit behind amyloid beta peptide related neurotoxicity in Alzheimer's disease: Oligomer size or conformation? *Alzheimers Res. Ther.* **2010**, *2*, 12. [CrossRef]
206. Malchiodi-Albedi, F.; Paradisi, S.; Matteucci, A.; Frank, C.; Diociaiuti, M. Amyloid oligomer neurotoxicity, calcium dysregulation, and lipid rafts. *Int. J. Alzheimers Dis.* **2011**, *2011*, 906964. [CrossRef]
207. Smith, L.M.; Strittmatter, S.M. Binding sites for amyloid-beta oligomers and synaptic toxicity. *Cold Spring Harb. Perspect. Med.* **2017**, *7*, a024075. [CrossRef]
208. Tanokashira, D.; Mamada, N.; Yamamoto, F.; Taniguchi, K.; Tamaoka, A.; Lakshmana, M.K.; Araki, W. The neurotoxicity of amyloid beta-protein oligomers is reversible in a primary neuron model. *Mol. Brain* **2017**, *10*, 4. [CrossRef]
209. Kepp, K.P.; Robakis, N.K.; Hoilund-Carlsen, P.F.; Sensi, S.L.; Vissel, B. The amyloid cascade hypothesis: An updated critical review. *Brain* **2023**, *146*, 3969–3990. [CrossRef]
210. Killingley, B.; Mann, A.J.; Kalinova, M.; Boyers, A.; Goonawardane, N.; Zhou, J.; Lindsell, K.; Hare, S.S.; Brown, J.; Frise, R.; et al. Safety, tolerability and viral kinetics during SARS-CoV-2 human challenge in young adults. *Nat. Med.* **2022**, *28*, 1031–1041. [CrossRef] [PubMed]
211. Appelman, B.; Charlton, B.T.; Goulding, R.P.; Kerkhoff, T.J.; Breedveld, E.A.; Noort, W.; Offringa, C.; Bloemers, F.W.; van Weeghel, M.; Schomakers, B.V.; et al. Muscle abnormalities worsen after postexertional malaise in long COVID. *Nat. Commun.* **2024**, *15*, 17. [CrossRef] [PubMed]
212. Vassiliou, A.G.; Jahaj, E.; Ilias, I.; Markaki, V.; Malachias, S.; Vrettou, C.; Ischaki, E.; Mastora, Z.; Douka, E.; Keskinidou, C.; et al. Lactate kinetics reflect organ dysfunction and are associated with adverse outcomes in intensive care unit patients with COVID-19 pneumonia: Preliminary results from a greek single-centre study. *Metabolites* **2020**, *10*, 386. [CrossRef]
213. Iepsen, U.W.; Plovsing, R.R.; Tjelle, K.; Foss, N.B.; Meyhoff, C.S.; Ryrsø, C.K.; Berg, R.M.G.; Secher, N.H. The role of lactate in sepsis and COVID-19: Perspective from contracting skeletal muscle metabolism. *Exp. Physiol.* **2021**, *107*, 665–673. [CrossRef]
214. Nechipurenko, Y.D.; Semyonov, D.A.; Lavrinenko, I.A.; Lagutkin, D.A.; Generalov, E.A.; Zaitceva, A.Y.; Matveeva, O.V.; Yegorov, Y.E. The role of acidosis in the pathogenesis of severe forms of COVID-19. *Biology* **2021**, *10*, 852. [CrossRef]
215. Carpenè, G.; Onorato, D.; Nocini, R.; Fortunato, G.; Rizk, J.G.; Henry, B.M.; Lippi, G. Blood lactate concentration in COVID-19: A systematic literature review. *Clin. Chem. Lab. Med.* **2022**, *60*, 332–337. [CrossRef] [PubMed]
216. de Boer, E.; Petrache, I.; Goldstein, N.M.; Olin, J.T.; Keith, R.C.; Modena, B.; Mohning, M.P.; Yunt, Z.X.; San-Millan, I.; Swigris, J.J. Decreased fatty acid oxidation and altered lactate production during exercise in patients with post-acute COVID-19 syndrome. *Am. J. Respir. Crit. Care Med.* **2022**, *205*, 126–129. [CrossRef] [PubMed]
217. van der Togt, V.; Rossman, J.S. Hypothesis: Inflammatory acid-base disruption underpins long COVID. *Front. Immunol.* **2023**, *14*, 1150105. [CrossRef] [PubMed]
218. Rutherford, G.; Manning, P.; Newton, J.L. Understanding muscle dysfunction in chronic fatigue syndrome. *J. Aging Res.* **2016**, *2016*, 2497348. [CrossRef] [PubMed]
219. Fernandez-Guerra, P.; Gonzalez-Ebsen, A.C.; Boonen, S.E.; Courraud, J.; Gregersen, N.; Mehlsen, J.; Palmfeldt, J.; Olsen, R.K.J.; Brinth, L.S. Bioenergetic and proteomic profiling of immune cells in myalgic encephalomyelitis/chronic fatigue syndrome patients: An exploratory study. *Biomolecules* **2021**, *11*, 961. [CrossRef] [PubMed]
220. Natelson, B.H.; Vu, D.; Coplan, J.D.; Mao, X.; Blate, M.; Kang, G.; Soto, E.; Kapusuz, T.; Shungu, D.C. Elevations of ventricular lactate levels occur in both chronic fatigue syndrome and fibromyalgia. *Fatigue* **2017**, *5*, 15–20. [CrossRef] [PubMed]
221. Murrough, J.W.; Mao, X.; Collins, K.A.; Kelly, C.; Andrade, G.; Nestadt, P.; Levine, S.M.; Mathew, S.J.; Shungu, D.C. Increased ventricular lactate in chronic fatigue syndrome measured by ^{1}H MRS imaging at 3.0 T. II: Comparison with major depressive disorder. *NMR Biomed.* **2010**, *23*, 643–650. [CrossRef] [PubMed]
222. Shungu, D.C.; Weiduschat, N.; Murrough, J.W.; Mao, X.; Pillemer, S.; Dyke, J.P.; Medow, M.S.; Natelson, B.H.; Stewart, J.M.; Mathew, S.J. Increased ventricular lactate in chronic fatigue syndrome. III. Relationships to cortical glutathione and clinical symptoms implicate oxidative stress in disorder pathophysiology. *NMR Biomed.* **2012**, *25*, 1073–1087. [CrossRef]

Journal of *Personalized Medicine*

MDPI

Article

The Presence of Ganglionic Acetylcholine Receptor Antibodies in Sera from Patients with Functional Gastrointestinal Disorders: A Preliminary Study

Shunya Nakane [1,2,3,*,†], Akihiro Mukaino [3,4,†], Yoshiaki Okumura [5], Hiroaki Hirosawa [1], Osamu Higuchi [2], Hidenori Matsuo [2], Mosaburo Kainuma [4] and Yuji Nakatsuji [1]

1 Department of Neurology, Faculty of Medicine, University of Toyama, Toyama 930-0194, Japan
2 Department of Neurology and Clinical Research, National Hospital Organization Nagasaki Kawatana Medical Center, Nagasaki 859-3615, Japan; matsuo.hidenori.wa@mail.hosp.go.jp (H.M.)
3 Department of Neurology, Graduate School of Medical Sciences, Kumamoto University, Kumamoto 860-8556, Japan
4 Department of Japanese Oriental Medicine, Graduate School of Medicine and Pharmaceutical Sciences, University of Toyama, Toyama 930-0194, Japan
5 Okumura Forest-All Clinic, Otsu 520-0831, Japan
* Correspondence: snakane@med.u-toyama.ac.jp
† These authors contributed equally to this work.

Abstract: Background: Functional gastrointestinal disorders (FGIDs), including functional dyspepsia (FD) and irritable bowel syndrome (IBS), are characterized by chronic and recurrent gastrointestinal symptoms. Clinically, FD and IBS often resemble gastrointestinal dysmotility caused by autoimmune autonomic neuropathy. We examined the seropositive frequency of autoantibodies against ganglionic nicotinic acetylcholine receptors (gnAChRs) in patients presenting with FGIDs. Objective: To elucidate the seropositivity of gnAChR antibodies and the clinical features of seropositive FD and IBS. Materials and Methods: We measured autoantibodies against the gnAChR α3 and β4subunits using luciferase immunoprecipitation systems. Serum samples from patients with any autonomic symptoms were obtained from hospitals in Japan between January 2012 and August 2018 (1787 serum samples of 1381 patients). We selected FD and IBS patients and compared the clinical characteristics and prevalence of autonomic symptoms between those with seropositive and seronegative IBS and FD. Results: Nine IBS and two FD cases (one comorbid case with IBS) were found. We found four patients (36.4%) in whom gnAChR antibodies were positive in these eleven patients. Sicca symptoms were observed in three of four cases (75%) of seropositive FGID compared with zero of seven cases (0%) of seronegative FGID. Conclusions: We found patients with gnAChR antibodies in FD and IBS patients. These data will be valuable for elucidating the pathophysiology of these FGIDs and developing new treatment strategies.

Keywords: functional gastrointestinal disorders; irritable bowel syndrome; functional dyspepsia; anti-ganglionic acetylcholine receptor; autoantibody

1. Introduction

Functional dyspepsia (FD) and irritable bowel syndrome (IBS) are the most common functional gastrointestinal disorders (FGIDs), affecting approximately 20% of the general population [1]. FGIDs are chronic or recurrent diseases in which abnormal bowel movements involving abdominal pain, diarrhea, and constipation, as well as gastric pain and early satiety, persist despite the absence of an organic disease on examination [2]. Similarly, autonomic neuropathy is a condition that presents functional impairment without organic abnormalities. We have previously performed clinical and basic studies of autoimmune autonomic ganglionopathy (AAG), in which autoantibodies against ganglionic nicotinic acetylcholine receptors (gnAChRs) are found in the serum of patients and play a key role in the pathogenesis of the disease [3,4]. Recently, the concept of autoimmune gastrointestinal dysmotility (AGID) has been proposed as a limited form of AAG [5,6]. AGID is becoming a broad concept that includes esophageal achalasia, diffuse esophageal spasm, gastroparesis, and intestinal pseudo-obstruction [7]. Although FD and IBS are clinically similar to the upper and lower gastrointestinal dysmotility in AAG or AGID, the relationship between the pathogenicity and gnAChR antibodies in FGIDs remains unresolved. Hence, we aimed to examine the seropositivity of gnAChR antibodies and the clinical characteristics of seropositive patients with FD and IBS.

2. Materials and Methods

2.1. Patient Cohort and Study Design

Our institution established the detection system of gnAChR α3 and β4 antibodies by a luciferase immunoprecipitation system (LIPS) assay in 2011, and since 2012, we have responded to requests for assays using serum samples from patients with any autonomic symptoms who visited a hospital in Japan [8,9]. We examined 1787 serum samples of 1381 patients with any autonomic symptoms who visited teaching and general hospitals throughout Japan between January 2012 and August 2018 [8,9]. The serum samples were centrifuged at 3000 rpm for 10 min and were then stored in cryovial tubes at −80 °C within 2 h of collection. The samples were later sent to Nagasaki Kawatana Medical Center or Kumamoto University Hospital.

2.2. Luciferase Immunoprecipitation System (LIPS) Assay for Anti-gAChR Abs

In the present study, we detected serum gnAChRα3 and gnAChRβ4 antibodies using the LIPS assay [9]. A National Institutes of Health group previously developed this efficient quantitative approach for the analysis of antibodies against human autoantigens in serum samples from patients [10,11]. We previously established and reported the use of the LIPS to diagnose AAG on the basis of IgGs to both α3 and β4 gnAChR subunits in serum samples from patients [9]. We measured the gnAChR antibodies at the Nagasaki Kawatana Medical Center and Kumamoto University Hospital, as previously described.

To generate luciferase reporters for the human gnAChR subunits, α3 and β4 (named gnAChRα3-GL and gnAChRβ4-GL, respectively) or full-length human gnAChRα3 (P32297; Promega Corporation, Madison, WI, USA) or gnAChRβ4 (P30296; Promega Corporation) were fused to a Gaussian luciferase (GL) mutant (GL^{8990}). Human embryonic kidney 293F cells (Life Technologies Corporation, Grand Island, NY, USA) were then transfected with an expression plasmid encoding either gnAChRα3-GL or gnAChRβ4-GL using FuGENE6 (Promega Corporation). The transfected cells were solubilized 2 days later using Tris-based saline containing 1% Triton™ X-100. To detect gnAChRα3 or gnAChRβ4 antibodies, 100 μL of the soluble fraction containing gnAChRα3-GL or gnAChRβ4-GL was incubated with 15 μL of human serum for 1 h at 4 °C. The fraction was then mixed with 15 μL of protein G-Sepharose (GE Healthcare, Little Chalfont, Buckinghamshire, UK) and 600 μL of phosphate-buffered saline with 3% bovine serum albumin and 0.05% Tween®-20, and the mixture was incubated for several hours at 4 °C. After centrifugation and two washes with phosphate-buffered saline containing 0.05% Tween®-20, the bioluminescence activity of the luciferase reporters in protein G-Sepharose was measured using a BioLux® GL Assay Kit (New England Biolabs, Ipswich, MA, USA) and a Lumat LB 9507 luminometer (Berthold Technologies GmbH & Co. KG, Bad Wildbad, Germany). The luminometer output was measured in relative luminescence units. Using anti-gnAChRα3 and anti-gnAChRβ4 antibody data from 73 healthy controls, cut-off values were calculated as the means + three standard deviations from the mean, as in a previous study. To evaluate the diagnostic accuracy of this LIPS assay, we verified the cut-off values for all data collected in previous studies [12,13]. Cut-off values for the sensitivity and specificity, as well as receiver operating characteristic curves, were obtained. According to these curves, we confirmed the most accurate cut-off values and calculated their sensitivity, specificity, and positive and negative predictive values. The area under the curve was 0.849 (95% confidence interval

[CI]: 0.786–0.911) for the LIPS assay of the anti-gnAChRα3 antibody. For an anti-gnAChRα3 antibody cut-off value of 1.0, the sensitivity and specificity values were 46.9% (95% CI: 33.7%–60.6%) and 99.2% (95% CI: 94.8%–100.0%), respectively, whereas the positive and negative predictive values were 95.8% and 81.8%, respectively. The area under the curve was 0.72 (95% CI: 0.632–0.807) for the LIPS assay of the anti-gnAChRβ4 antibody. For an anti-gnAChRβ4 antibody cut-off value of 1.0, the sensitivity and specificity values were 14.3% (95% CI: 6.9%–27.1%) and 100.0% (95% CI: 96.2%–100.0%), respectively, whereas the positive and negative predictive values were 100.0% and 74.4%, respectively.

The antibody levels were expressed as an antibody index (AI), which was calculated as follows: AI = (measured value in the serum sample [in relative luminescence units])/(cut-off value [in relative luminescence units]). The normal AI value, established based on data from healthy individuals, is <1.0. The diagnosis of FD and IBS was confirmed using the Rome IV diagnostic guidelines [14,15]. Clinical diagnoses were performed in each hospital. We compared the clinical data and the prevalence of autonomic symptoms between seropositive and seronegative FD and IBS patients.

2.3. Ethical Approval

All patients provided written informed consent for the storage and use of their serum and clinical information for research purposes. The Human Ethics Committees at the Nagasaki Kawatana Medical Center and Kumamoto University Hospital (Japan) approved this study (approval numbers 2011-21 and 1281, respectively).

2.4. Clinical Assessment

Specific questionnaires and consent forms were sent to the physicians who referred us to patients, and the data were sorted and analyzed. The questionnaire consisted of six categories with the following entries: (1) age, sex, clinical diagnosis, age at onset of disease, antecedent infection, and mode of symptom onset; (2) autonomic manifestations described below; (3) extra-autonomic manifestations (sensory disturbance, motor symptoms, deep tendon reflexes, gait, and other neurological findings); (4) comorbid diseases (endocrine disorders, tumors, and autoimmune diseases); (5) autonomic testing; (6) other laboratory findings. Regarding the mode of symptom onset, acute onset and subacute onset were defined as reaching peak autonomic symptoms within 3 months. Chronic onset was defined as reaching peak autonomic symptoms after 3 months.

We determined the presence or absence of the following functions controlled by the autonomic nervous system, as reported in our previous study: syncope or orthostatic hypotension and orthostatic intolerance; arrhythmia; pupillary dysfunction; sicca complex; coughing episodes; skin dryness or hypohidrosis/anhidrosis indicating heat intolerance; upper gastrointestinal system problems; diarrhea or constipation indicating dysfunction of the lower gastrointestinal system; dysuria or urinary retention needing catheterization for bladder dysfunction; and sexual dysfunction [8]. We selected patients who were diagnosed with FD or IBS from the patient cohort and divided them into gnAChR antibody-positive and gnAChR antibody-negative groups for a comparative analysis of their clinical features.

2.5. COMPASS

Patients with FD or IBS enrolled after April 2014 completed a self-administered questionnaire. COMPASS is a shortened version of the Composite Autonomic Symptom Score and was designed to quantitatively assess autonomic symptoms [16]. It has six subscale weighted scores in the following domains: orthostatic intolerance (four items; range, 0–40), vasomotor (three items; range, 0–5), secretomotor (four items; range, 0–15), gastrointestinal (12 items; range, 0–25), bladder (three items; range, 0–10), and pupillomotor (five items; range, 0–5). The COMPASS assessment is weighted according to published scoring methods to yield a total score of 0–100, with a score of 100 representing the highest, most severe degree of autonomic symptom burden. The mean $\pm$ standard deviation score in healthy control subjects for this questionnaire was reported as 9.67 ± 8.1 [17]. In the present study, 11 subjects completed the Japanese language version of the questionnaire within 15 min. We excluded questions related to the vasomotor and pupillomotor domains because it is occasionally difficult for Japanese people to judge color changes of the skin on an individual basis, and it is not the custom for middle-aged and older persons to wear sunglasses or tinted glasses in Japan. The total scores were calculated by the summation of the individual item scores, with a possible maximum score of 90 [8,18].

3. Results

In this study, we identified two patients with FD and nine patients with IBS in 1381 patients. One of the former patients also had IBS. Among those 11 patients, 4 patients had gnAChR antibodies. Of the four patients with gnAChR antibody-positive FGID, three had IBS only, and one patient had coexistent FD and IBS. Single seropositivity for gnAChR α3 antibodies was observed in two patients, while single seropositivity for gnAChR β4 antibodies was not observed. Two of the four samples were positive for both gnAChR α3 and β4 antibodies. One of these double-positive patients was particularly refractory to the clinical manifestations of IBS.

The clinical features of the four patients in the gnAChR antibody-positive group and the seven patients in the antibody-negative group were compared, as shown in Table 1. Many of the items that were compared, including the usual epidemiological items and each autonomic symptom, as well as the COMPASS total and domain-specific scores (Table 2), showed no significant differences. Only sicca symptoms were significantly more frequent in the antibody-positive group (75% vs. 0%, $p = 0.042$).

Table 1. Clinical profiles of patients with FGID in the presence or absence of gAChR Abs.

Characteristics	FGID with gAChR Abs (n = 4)	FGID without gAChR Abs (n = 7)	*p* Value
Age (average, years)	71.5	59.7	0.262
Sex, female (%)	3 (75.0)	5 (71.3)	0.166
Mode of onset, chronic (%)	4 (100.0)	6 (85.7)	0.788
Antecedent infection (%)	0 (0.0)	1 (14.3)	0.788
Orthostatic hypotension (%)	0 (0.0)	2 (28.6)	0.527
Orthostatic intolerance (%)	1 (25.0)	4 (57.1)	0.412
Arrhythmia (%)	0 (0.0)	1 (14.3)	0.788
Pupillary abnormality (%)	0 (0.0)	0 (0.0)	1.000
Sicca (%)	3 (75.0)	0 (0.0)	0.042
Coughing episode (%)	0 (0.0)	1 (14.3)	0.788
Anhidrosis (%)	1 (25.0)	0 (0.0)	0.788
Upper GI dysfunction (%)	4 (100.0)	7 (100.0)	1.000
Lower GI dysfunction (%)	4 (100.0)	7 (100.0)	1.000
Bladder dysfunction (%)	2 (50.0)	6 (85.7)	0.412
Sexual dysfunction (%)	0 (0.0)	0 (0.0)	1.000

$p < 0.05$ was considered statistically significant. Abbreviations: FGID = functional gastrointestinal disorders; gAChR = ganglionic acetylcholine receptor; Abs = autoantibodies; GI = gastrointestinal.

Table 2. Comparison of the COMPASS of FGID patients with and without gAChR Abs.

Characteristics	FGID with gAChR Abs (n = 4)	FGID without gAChR Abs (n = 7)	*p* Value
COMPASS total score (average)	16.5	20.8	0.455
COMPASS orthostatic intolerance score (average)	3.0	8.7	0.294
COMPASS secretomotor score (%)	2.6	0.4	0.114

Table 2. *Cont.*

Characteristics	FGID with gAChR Abs (n = 4)	FGID without gAChR Abs (n = 7)	*p* Value
COMPASS gastrointestinal score	8.6	8.6	0.975
COMPASS bladder score	1.4	3.1	0.185

Abbreviations: COMPASS = Composite Autonomic Symptom Score; FGID = functional gastrointestinal disorders; gAChR = ganglionic acetylcholine receptor; Abs = autoantibodies.

Illustrative Cases

Patient 1. An 85-year-old man had been affected by IBS for at least 5 years and experienced abdominal pain, constipation, and abdominal bloating soon after eating. The patient did not have orthostatic hypotension/intolerance and had objective findings of dry mouth and upper and lower gastrointestinal dysmotility. The COMPASS also reflected these clinical findings. The patient had only autoantibodies against gnAChRα3, and the serum levels of the gnAChR autoantibodies were 1.017 antibody index (A.I.) (α3) and 0.641 A.I. (β4).

Patient 2. A 51-year-old woman had been affected by FD and IBS for at least 5 years and had nausea, vomiting, and diarrhea. She visited many hospitals and received multiple antiemetic prescriptions, but the nausea remained. She took anti-diarrheal medicines year-round for chronic diarrhea that was triggered by more intense diarrhea that occurred when she consumed high-fat meals. She also had severe left back pain after eating, which led her previous healthcare providers to suspect chronic pancreatitis, and she underwent endoscopic ultrasonography and other tests that did not reveal any abnormalities. An overactive bladder had also been diagnosed, and incontinence was a rare occurrence because of an inability to hold back urine. The patient had autoantibodies for both gnAChRα3 and β4, and her serum levels of the gnAChR autoantibodies were 2.218 A.I. (α3) and 1.135 A.I. (β4).

Patient 3. A 75-year-old woman had been affected by IBS for several years and experienced constipation, appetite loss, nausea, and vomiting. She complained of numbness in both lower extremities and pain in her buttocks in addition to the symptoms attributed to gastrointestinal dysmotility. Seeking further examination and treatment for these symptoms, she consulted a neurologist as well as a local orthopedic surgeon and a pain clinic and was eventually diagnosed with fibromyalgia on the basis of these symptoms. The patient consistently complained of bloating, which was exacerbated by drugs used to treat fibromyalgia, which caused nausea, making it difficult for us to treat both the gastrointestinal dysmotility and fibromyalgia. The patient had autoantibodies for both gnAChRα3 and β4, and her serum levels of the gnAChR autoantibodies were 1.444 A.I. (α3) and 1.078 A.I. (β4).

Patient 4. A 75-year-old woman had been affected by IBS for at least 2 years and experienced abdominal pain, diarrhea, constipation, and abdominal bloating soon after eating. The patient had abdominal pain and discomfort and constipation for several days, followed by diarrhea for several days; therefore, constipation and diarrhea appeared alternately. During the constipation, her abdomen was tense and painful. We prescribed various medications, including Chinese herbal medicines, for the constipation and diarrhea. The patient sometimes had sudden diarrhea when she was away from home; hence, she was afraid to go out. The patient had only autoantibodies against gnAChRα3, and her serum levels for gnAChR autoantibodies were 1.214 A.I. (α3) and 0.175 A.I. (β4).

4. Discussion

This preliminary study yielded two findings. First, some of the patients diagnosed with FGID had gnAChR antibodies, and second, the clinical symptom sicca complex was frequently observed in gnAChR antibody-positive FGID patients. These results raise the question of whether gnAChR antibody positivity is involved in the pathogenesis of FGID patients or is coincidental. However, further issues remain to be explored in future research.

Although 4 of the 11 patients with FGID in the current study were positive for gnAChR antibodies, it was difficult to determine whether they showed a consistent trend based on age or sex. It was also difficult to determine whether higher levels of gnAChR antibodies or autoantibody positivity regarding both subunits were associated with the severity of abdominal symptoms.

Gastrointestinal symptoms are common and highly prevalent. However, not all patients presenting with gastrointestinal symptoms have a specific organic etiology. Some cases involve FGIDs, such as FD or IBS, in which patients with these conditions suffer from chronic and fluctuating symptoms. The pathogenesis of FGIDs has been described as a gut–brain interaction disorder, which has been studied regarding a variety of aspects, including movement disorders, visceral hypersensitivity, altered mucosal and immune function, altered gut microbiota, and altered central nervous system processing [2,15]. Because the results of the present study suggest the presence of some antibodies in FGID patients, we must consider the possibility that the pathology is mediated by dysimmune status. Atopic and autoimmune diseases, independent of psychological distress, were reported to be risk factors for FGIDs, with psoriasis and rheumatoid arthritis in particular being independent risk factors for IBS [19]. Rheumatoid arthritis was also significantly associated with IBS in another survey of 850 pairs of Swedish twins aged 18–85 years [20]. The involvement of neuro-immune interactions in FGID has been discussed previously [21–24], and various basic studies have been conducted on the involvement of specific autoantibodies, showing positive and negative results regarding their presence [25–29]. Further detailed clinical and basic studies are needed to determine whether the gnAChR antibodies found in this preliminary study are pathogenic autoantibodies in FGID patients.

Here, we present a relatively new disease concept, autoimmune gastrointestinal dysmotility, a limited form of AAG, which is an autonomic disorder in which gastrointestinal motility disorders are in the foreground of the clinical presentation [5–7]. Previous reports have indicated that gastrointestinal dysmotility, such as constipation, diarrhea, alternating constipation–diarrhea, and ileus, as well as orthostatic hypotension and orthostatic intolerance, occur frequently in AAG patients who test positive for gnAChR antibodies, a known pathogenic autoantibody in AAG [7,18]. Although its name implies a localized condition, this disease can present with varying degrees of symptoms from other autonomic domains [7,18]. Another clinical feature of AAG is the presentation of extra-autonomic manifestations, including a tendency to coexist with autoimmune diseases such as autoimmune rheumatic diseases and autoimmune thyroid diseases [8]. Interestingly, complications of gastrointestinal dysmotility in autoimmune rheumatic diseases, such as Sjögren's syndrome, have been frequently presented in practice [7,30]. Because the present study showed a significantly increased frequency of the sicca complex in patients with gnAChR antibody-positive FGID, we must consider the possibility that these cases were actually autoimmune gastrointestinal dysmotility, a limited form of AAG.

In the peripheral autonomic ganglia, nAChRs (equivalent to gnAChRs) are expressed by neurons in sympathetic, parasympathetic, and enteric ganglia. Patients with AAG often harbor autoantibodies against gAChRs, which may disrupt synaptic transmission in autonomic ganglia and lead to autonomic failure [31,32]. However, the pathogenicity of the gnAChR antibodies, i.e., how the autoantibodies cause autonomic dysfunction, is not entirely clear. More recently, autoantibodies targeting neurotransmitter receptors and related proteins have emerged as an often severe but treatable cause of neurological disease [33]. Autoantibodies against nAChRs in autonomic ganglia should be considered similar to autoantibodies against neurotransmitter receptors when discussing pathogenesis, although AAG is also often present in refractory cases. Previous animal model studies of AAG and an in vitro study using the nAChRα3 subunit expressed in human embryonic kidney cells have shown that autoantibodies for the nAChR subunit cross-link and internalize the postsynaptic nAChR, leading to its degradation, which was shown to be the pathogenic mechanism [34–37]. Animal models of gastrointestinal hypomotility have previously been established by the intraperitoneal injection of live nAChRα3-expressing xenogeneic cells [34]. Blue dye passaging, radiochemical, and immunohistochemical evaluations demonstrated the small intestinal transit of these cells, indicating that high concentrations of nAChRα3-IgG in serum are required for intestinal nAChRα3 depletion. In addition, no loss of ganglion neurons was observed. Recently, we developed a novel murine model of autoimmune dysautonomia by nAChRα3 immunization and identified two key immunogenic peptides that could effectively prime helper T cells [38]. Physiological testing confirmed the delayed intestinal transit in these active immunized mice, and ileus occurred because of intestinal accumulation in one of the mice. It is conceivable that gnAChR antibodies act functionally on the receptors. However, further investigation is required to determine whether these antibodies have agonistic or antagonistic effects on the receptor [39]. It remains unclear how gnAChR autoantibodies are involved in gastrointestinal dysmotility in the enteric nervous system [7]. According to the results of the present study, it is clear that an important question to be resolved in the future is whether some FGIDs overlap with the AGID concept [7]. AGID is a condition that can occur at each level of the gastrointestinal tract, including the esophagus, stomach, small intestine, large intestine, rectum, or anus, and diseases at each site include achalasia, diffuse esophageal spasm, gastroparesis, pyloric stenosis, intestinal pseudo-obstruction, slow intestinal transit, colonic inertia, and anal spasm [7]. Various autoantibodies have been implicated in AGID. Moreover, in addition to gnAChR antibodies, muscle nicotinic AChR, voltage-gated potassium channel complex, voltage-gated calcium channel (P/Q type and N type), and glutamic acid decarboxylase antibodies have been previously reported [5,7,40,41]. The involvement of these autoantibodies that target the enteric nervous system in paraneoplastic neurologic syndromes including Lambert–Eaton syndrome, Chagas disease, and diabetes, leading to impaired gastrointestinal motility, has also been reported [42–44]. Based on previous reports, it is difficult to determine whether AGID is simply a limited form of AAG and a clinically heterogeneous group [7]. The present study indicates that some cases of FGID may be AGID. The finding that some patients with FGID have autoantibodies is important in considering the pathogenesis of FGID and, ultimately, its treatment. These antineuronal autoantibodies have been found to be present in AGIDs at each level of the gastrointestinal tract, and it is possible that some FGIDs also have antibody-positive cases. Therefore, the antineuronal autoantibodies listed here should be measured in future large studies.

This study has several limitations. It is preliminary and is an observational study with a small sample size, albeit with expert clinical diagnosis. It is necessary to confirm in the future whether gnAChR antibodies are present in a greater number of FGID patients in a prospective multicenter study. Furthermore, the LIPS assay we established was used to detect gnAChR antibodies. Recently, other new antibody assays such as flow cytometry and cell-based assays have been reported [45,46]. In addition to conventional detection methods such as radioimmunoprecipitation and LIPS assays, it is necessary to verify the presence of autoantibodies using different detection systems that incorporate other new methods. After such validation, the role of gnAChR antibodies in the pathogenesis of patients with FGID should be clarified. This will allow us to understand the true pathogenic role of this autoantibody and will provide an opportunity to investigate the possibility of immunotherapy for antibody-induced autonomic dysfunction.

In summary, we reported the presence of gnAChR antibody-positive cases of FD and IBS. These cases will be valuable in elucidating the pathophysiology of these FGIDs and developing new treatment strategies.

Author Contributions: Conceptualization, S.N.; Methodology, A.M. and O.H.; Formal analysis, S.N.; Investigation, S.N., A.M., Y.O. and H.M.; Writing—original draft, S.N., A.M. and H.H.; Writing—review and editing, S.N., H.M., M.K. and Y.N.; Funding acquisition, S.N. All authors have read and agreed to the published version of the manuscript.

Funding: This study was supported by the Ministry of Education, Culture, Sports, Science, and Technology of Japan JSPS KAKENHI (Grant Numbers 19K17012 and 19H03549). This study was also supported in part by the Health and Labour Sciences Research Grant on Rare and Intractable Diseases from the Ministry of Health, Labour, and Welfare of Japan.

Institutional Review Board Statement: This study was approved by the Human Ethics Committees of Nagasaki Kawatana Medical Center on 9 February 2012 and Kumamoto University Hospital on 27 December 2016 (approval numbers 2011-21 and 1281, respectively).

Informed Consent Statement: All patients provided written informed consent for the storage and use of their serum and clinical information for research purposes.

Data Availability Statement: The data presented in this study are available on request from the corresponding author.

Conflicts of Interest: The authors declare that the research was conducted in the absence of any commercial or financial relationships that could be construed as a potential conflict of interest.

References

1. Koloski, N.A.; Talley, N.J.; Boyce, P.M. Epidemiology and health care seeking in the functional GI disorders: A population-based study. *Am. J. Gastroenterol.* **2002**, *97*, 2290–2299. [CrossRef] [PubMed]
2. Black, C.J.; Drossman, D.A.; Talley, N.J.; Ruddy, J.; Ford, A.C. Functional Gastrointestinal Disorders: Advances in Understanding and Management. *Lancet* **2020**, *396*, 1664–1674. [CrossRef] [PubMed]
3. Vernino, S.; Low, P.A.; Fealey, R.D.; Stewart, J.D.; Farrugia, G.; Lennon, V.A. Autoantibodies to Ganglionic Acetylcholine Receptors in Autoimmune Autonomic Neuropathies. *N. Engl. J. Med.* **2000**, *343*, 847–855. [CrossRef] [PubMed]
4. Wang, Z.; Low, P.A.; Jordan, J.; Freeman, R.; Gibbons, C.H.; Schroeder, C.; Sandroni, P.; Vernino, S. Autoimmune autonomic ganglionopathy: IgG effects on ganglionic acetylcholine receptor current. *Neurology* **2007**, *68*, 1917–1921. [CrossRef] [PubMed]
5. Dhamija, R.; Tan, K.M.; Pittock, S.J.; Foxx-Orenstein, A.; Benarroch, E.; Lennon, V.A. Serologic profiles aiding the diagnosis of autoimmune gastrointestinal dysmotility. *Clin. Gastroenterol. Hepatol.* **2008**, *6*, 988–992. [CrossRef] [PubMed]
6. Flanagan, E.P.; Saito, Y.A.; Lennon, V.A.; McKeon, A.; Fealey, R.D.; Szarka, L.A.; Murray, J.A.; Foxx-Orenstein, A.E.; Fox, J.C.; Pittock, S.J. Immunotherapy trial as diagnostic test in evaluating patients with presumed autoimmune gastrointestinal dysmotility. *Neurogastroenterol. Motil.* **2014**, *26*, 1285–1297. [CrossRef] [PubMed]
7. Nakane, S.; Mukaino, A.; Ihara, E.; Ogawa, Y. Autoimmune gastrointestinal dysmotility: The interface between clinical immunology and neurogastroenterology. *Immunol. Med.* **2021**, *44*, 74–85. [CrossRef] [PubMed]
8. Nakane, S.; Mukaino, A.; Higuchi, O.; Yasuhiro, M.; Takamatsu, K.; Yamakawa, M.; Watari, M.; Tawara, N.; Nakahara, K.-I.; Kawakami, A.; et al. A comprehensive analysis of the clinical characteristics and laboratory features in 179 patients with autoimmune autonomic ganglionopathy. *J. Autoimmun.* **2020**, *108*, 102403. [CrossRef] [PubMed]
9. Nakane, S.; Higuchi, O.; Koga, M.; Kanda, T.; Murata, K.; Suzuki, T.; Kurono, H.; Kunimoto, M.; Kaida, K.-I.; Mukaino, A.; et al. Clinical features of autoimmune autonomic ganglionopathy and the detection of subunit-specific autoantibodies to the ganglionic acetylcholine receptor in Japanese patients. *PLoS ONE* **2015**, *10*, e0118312. [CrossRef]
10. Burbelo, P.D.; Ching, K.H.; Klimavicz, C.M.; Iadarola, M.J. Antibody profiling by Luciferase Immunoprecipitation Systems (LIPS). *J. Vis. Exp.* **2009**, *32*, 1549.
11. Burbelo, P.D.; Lebovitz, E.E.; Notkins, A.L. Luciferase immunoprecipitation systems for measuring antibodies in autoimmune and infectious diseases. *Transl. Res.* **2015**, *165*, 325–335. [CrossRef] [PubMed]
12. Maeda, Y.; Migita, K.; Higuchi, O.; Mukaino, A.; Furukawa, H.; Komori, A.; Nakamura, M.; Hashimoto, S.; Nagaoka, S.; Abiru, S.; et al. Association between Anti-Ganglionic Nicotinic Acetylcholine Receptor (gAChR) Antibodies and HLA-DRB1 Alleles in the Japanese Population. *PLoS ONE* **2016**, *11*, e0146048. [CrossRef] [PubMed]
13. Nakane, S.; Mukaino, A.; Higuchi, O.; Watari, M.; Maeda, Y.; Yamakawa, M.; Nakahara, K.; Takamatsu, K.; Matsuo, H.; Ando, Y. Autoimmune autonomic ganglionopathy: An update on diagnosis and treatment. *Expert Rev. Neurother.* **2018**, *18*, 953–965. [CrossRef] [PubMed]
14. Mearin, F.; Lacy, B.E.; Chang, L.; Chey, W.D.; Lembo, A.J.; Simren, M.; Spiller, R. Bowel Disorders. *Gastroenterology* **2016**, *150*, 1393–1407.
15. Drossman, D.A.; Hasler, W.L. Rome IV—Functional GI Disorders: Disorders of Gut-Brain Interaction. *Gastroenterology* **2016**, *150*, 1257–1261. [CrossRef] [PubMed]
16. Sletten, D.M.; Suarez, G.A.; Low, P.A.; Mandrekar, J.; Singer, W. COMPASS 31: A Refined and Abbreviated Composite Autonomic Symptom Score. *Mayo Clin. Proc.* **2012**, *87*, 1196–1201. [CrossRef] [PubMed]
17. Cortez, M.; Reddy, S.N.; Goodman, B.; Carter, J.; Wingerchuk, D. Autonomic symptom burden is associated with MS-related fatigue and quality of life. *Mult. Scler. Relat. Disord.* **2015**, *4*, 258–263. [CrossRef] [PubMed]
18. Mukaino, A.; Minami, H.; Isomoto, H.; Hamamoto, H.; Ihara, E.; Maeda, Y.; Higuchi, O.; Okanishi, T.; Kokudo, Y.; Deguchi, K.; et al. Anti-ganglionic AChR antibodies in Japanese patients with motility disorders. *J. Gastroenterol.* **2018**, *53*, 1227–1240. [CrossRef] [PubMed]

19. Koloski, N.; Jones, M.; Walker, M.M.; Veysey, M.; Zala, A.; Keely, S.; Holtmann, G.; Talley, N.J. Population based study: Atopy and autoimmune diseases are associated with functional dyspepsia and irritable bowel syndrome, independent of psychological distress. *Aliment. Pharmacol. Ther.* **2019**, *49*, 546–555. [CrossRef]
20. Svedberg, P.; Johansson, S.; Wallander, M.; Hamelin, B.; Pedersen, N.L. Extra-intestinal manifestations associated with irritable bowel syndrome: A twin study. *Aliment. Pharmacol. Ther.* **2002**, *16*, 975–983. [CrossRef]
21. De Giorgio, R.; Guerrini, S.; Barbara, G.; Stanghellini, V.; De Ponti, F.; Corinaldesi, R.; Moses, P.L.; Sharkey, K.A.; Mawe, G.M. Inflammatory neuropathies of the enteric nervous system. *Gastroenterology* **2004**, *126*, 1872–1883. [CrossRef] [PubMed]
22. Kraneveld, A.D.; Rijnierse, A.; Nijkamp, F.P.; Garssen, J. Neuro-immune interactions in inflammatory bowel disease and irritable bowel syndrome: Future therapeutic targets. *Eur. J. Pharmacol.* **2008**, *585*, 361–374. [CrossRef] [PubMed]
23. Burns, G.; Carroll, G.; Mathe, A.; Horvat, J.; Foster, P.; Walker, M.M.; Talley, N.J.; Keely, S. Evidence for Local and Systemic Immune Activation in Functional Dyspepsia and the Irritable Bowel Syndrome: A Systematic Review. *Am. J. Gastroenterol.* **2019**, *114*, 429–436. [CrossRef] [PubMed]
24. Burns, G.L.; Bruce, J.K.; Minahan, K.; Mathe, A.; Fairlie, T.; Cameron, R.; Naudin, C.; Nair, P.M.; Potter, M.D.E.; Irani, M.Z.; et al. Type 2 and type 17 effector cells are increased in the duodenal mucosa but not peripheral blood of patients with functional dyspepsia. *Front. Immunol.* **2023**, *13*, 1051632. [CrossRef] [PubMed]
25. De Giorgio, R.; Bovara, M.; Barbara, G.; Canossa, M.; Sarnelli, G.; De Ponti, F.; Stanghellini, V.; Tonini, M.; Cappello, S.; Pagnotta, E.; et al. Anti-HuD-induced neuronal apoptosis underlying paraneoplastic gut dysmotility. *Gastroenterology* **2003**, *125*, 70–79. [CrossRef] [PubMed]
26. Pittock, S.J.; Lennon, V.A.; Dege, C.L.; Talley, N.J.; Locke, G.R., 3rd. Neural autoantibody evaluation in functional gastrointestinal disorders: A population-based case-control study. *Dig. Dis. Sci.* **2011**, *56*, 1452–1459. [CrossRef] [PubMed]
27. Fan, W.; Fei, G.; Li, X.; Wang, X.; Hu, C.; Xin, H.; Sun, X.; Li, Y.; Wood, J.D.; Fang, X. Sera with anti-enteric neuronal antibodies from patients with irritable bowel syndrome promote apoptosis in myenteric neurons of guinea pigs and human SH-Sy5Y cells. *Neurogastroenterol. Motil.* **2018**, *30*, e13457. [CrossRef] [PubMed]
28. Fan, W.; Fang, X.; Hu, C.; Fei, G.; Xiao, Q.; Li, Y.; Li, X.; Wood, J.D.; Zhang, X. Multiple rather than specific autoantibodies were identified in irritable bowel syndrome with HuProt™ proteome microarray. *Front. Physiol.* **2022**, *13*, 1010069. [CrossRef] [PubMed]
29. Bierła, J.B.; Cukrowska, B.; Skrzydło-Radomańska, B.; Prozorow-Król, B.; Kurzeja-Mirosław, A.; Cichoż-Lach, H.; Laskowska, K.; Sowińska, A.; Majsiak, E. The Occurrence of Gluten-Related Antibodies, Sensitization to Selected Food Allergens, and Antibodies against Intrinsic Factor in Adult Patients with Diarrhea-Predominant Irritable Bowel Syndrome. *J. Pers. Med.* **2023**, *13*, 1165. [CrossRef]
30. Volter, F.; Fain, O.; Mathieu, E.; Thomas, M. Esophageal function and Sjögren's syndrome. *Dig. Dis. Sci.* **2004**, *49*, 248–253. [CrossRef]
31. Vernino, S.; Lindstrom, J.; Hopkins, S.; Wang, Z.; Low, P.A.; Muscle Study Group. Characterization of ganglionic acetylcholine receptor autoantibodies. *J. Neuroimmunol.* **2008**, *197*, 63–69. [CrossRef] [PubMed]
32. Wang, Z.; Low, P.A.; Vernino, S. Antibody-mediated impairment and homeostatic plasticity of autonomic ganglionic synaptic transmission. *Exp. Neurol.* **2010**, *222*, 114–119. [CrossRef] [PubMed]
33. Crisp, S.J.; Kullmann, D.M.; Vincent, A. Autoimmune synaptopathies. *Nat. Rev. Neurosci.* **2016**, *17*, 103–117. [CrossRef] [PubMed]
34. Meeusen, J.W.; Haselkorn, K.E.; Fryer, J.P.; Kryzer, T.J.; Gibbons, S.J.; Xiao, Y.; Lennon, V.A. Gastrointestinal hypomotility with loss of enteric nicotinic acetylcholine receptors: Active immunization model in mice. *Neurogastroenterol. Motil.* **2013**, *25*, 84–88. [CrossRef] [PubMed]
35. Lennon, V.A.; Ermilov, L.G.; Szurszewski, J.H.; Vernino, S. Immunization with neuronal nicotinic acetylcholine receptor induces neurological autoimmune disease. *J. Clin. Invest.* **2003**, *111*, 907–913. [CrossRef] [PubMed]
36. Vernino, S.; Low, P.A.; Lennon, V.A. Experimental Autoimmune Autonomic Neuropathy. *J. Neurophysiol.* **2003**, *90*, 2053–2059. [CrossRef] [PubMed]
37. Kobayashi, S.; Yokoyama, S.; Maruta, T.; Negami, M.; Muroyama, A.; Mitsumoto, Y.; Iwasa, K.; Yamada, M.; Yoshikawa, H. Autoantibody-induced internalization of nicotinic acetylcholine receptor α3 subunit exogenously expressed in human embryonic kidney cells. *J. Neuroimmunol.* **2013**, *257*, 102–106. [CrossRef] [PubMed]
38. Yamakawa, M.; Nakane, S.; Ihara, E.; Tawara, N.; Ikeda, H.; Igarashi, Y.; Komohara, Y.; Takamatsu, K.; Ikeda, T.; Tomita, Y.; et al. A novel murine model of autoimmune dysautonomia by α3 nicotinic acetylcholine receptor immunization. *Front. Neurosci.* **2022**, *16*, 1006923. [CrossRef] [PubMed]
39. Nakane, S.; Yamakawa, M.; Nakatsuji, Y. Autoantibodies against gAChR in humans and mouse models of autoimmune autonomic ganglionopathy. *Clin. Exp. Neuroimmunol.* **2024**, *15*, 16–23. [CrossRef]
40. Hubball, A.W.; Lang, B.; Souza, M.A.; Curran, O.D.; Martin, J.E.; Knowles, C.H. Voltage-gated potassium channel (K(v) 1) autoantibodies in patients with chagasic gut dysmotility and distribution of K(v) 1 channels in human enteric neuromusculature (autoantibodies in GI dysmotility). *Neurogastroenterol. Motil.* **2012**, *24*, 719–728. [CrossRef]
41. McMillan, H.J.; Srinivasan, J. Achalasia, chronic sensory neuropathy, and N-type calcium channel autoantibodies: Beneficial response to IVIG. *Clin. J. Gastroenterol.* **2010**, *3*, 78–82. [CrossRef] [PubMed]
42. Hirano, I.; Pandolfino, J. Chronic intestinal pseudo-obstruction. *Dig. Dis.* **2000**, *18*, 83–92. [CrossRef] [PubMed]

43. Srinivasan, S.; Wiley, J.W. New insights into neural injury, repair, and adaptation in visceral afferents and the enteric nervous system. *Curr. Opin. Gastroenterol.* **2000**, *16*, 78. [CrossRef]
44. Goin, J.C.; Sterin–Borda, L.; Bilder, C.R.; Varrica, L.M.; Iantorno, G.; Ríos, M.C.; Borda, E. Functional implications of circulating muscarinic cholinergic receptor autoantibodies in chagasic patients with achalasia. *Gastroenterology* **1999**, *117*, 798–805. [CrossRef] [PubMed]
45. Urriola, N.; Spies, J.M.; Blazek, K.; Lang, B.; Adelstein, S. A Flow Cytometric Assay to Detect Functional Ganglionic Acetylcholine Receptor Antibodies by Immunomodulation in Autoimmune Autonomic Ganglionopathy. *Front. Immunol.* **2021**, *12*, 705292. [CrossRef]
46. Karagiorgou, K.; Dandoulaki, M.; Mantegazza, R.; Andreetta, F.; Furlan, R.; Lindstrom, J.; Zisimopoulou, P.; Chroni, E.; Kokotis, P.; Anagnostou, E.; et al. Novel Cell-Based Assay for Alpha-3 Nicotinic Receptor Antibodies Detects Antibodies Exclusively in Autoimmune Autonomic Ganglionopathy. *Neurol.—Neuroimmunol. Neuroinflammation* **2022**, *9*, 1162. [CrossRef]

Journal of
*Personalized
Medicine*

Review

Autoimmunity in Syndromes of Orthostatic Intolerance: An Updated Review

Clarissa Pena [1,*], Abdelmoniem Moustafa [2], Abdel-Rhman Mohamed [1] and Blair Grubb [2]

1 Department of Internal Medicine, University of Toledo, Toledo, OH 43614, USA; abdelrhman.mohamed@utoledo.edu
2 Division of Cardiovascular Medicine, University of Toledo, Toledo, OH 43614, USA; abdelmoniem.moustafa@utoledo.edu (A.M.); blair.grubb@utoledo.edu (B.G.)
* Correspondence: clarissa.pena@utoledo.edu; Tel.: +1-(419)-383-3685

Abstract: Orthostatic intolerance is a broad term that represents a spectrum of dysautonomic disorders, including postural orthostatic tachycardia syndrome (POTS) and orthostatic hypotension (OH), as manifestations of severe autonomic failure. While the etiology of orthostatic intolerance has not yet fully been uncovered, it has been associated with multiple underlying pathological processes, including peripheral neuropathy, altered renin–aldosterone levels, hypovolemia, and autoimmune processes. Studies have implicated adrenergic, cholinergic, and angiotensin II type I autoantibodies in the pathogenesis of orthostatic intolerance. Several case series have demonstrated that immunomodulation therapy resulted in favorable outcomes, improving autonomic symptoms in POTS and OH. In this review, we highlight the contemporary literature detailing the association of autoimmunity with POTS and OH.

Keywords: adrenergic antibodies; cholinergic antibodies; autonomic dysregulation; POTS; angiotensin II type I antibodies; COVID-19

1. Introduction

Orthostatic intolerance is a broad term that represents a spectrum of dysautonomic disorders, including postural orthostatic tachycardia syndrome (POTS) and orthostatic hypotension (OH). The hallmark of orthostatic intolerance is the triggering of symptoms upon standing [1,2].

OH is defined as a decrease in blood pressure of 20 mm Hg for systolic blood pressure or 10 mm Hg for diastolic blood pressure within 3 min of standing [3]. Older age has been associated with the development of OH [4]. POTS is defined as being in an upright posture with a sustained increase in heart rate of 30 beats/minute within 10 min of standing or head-up tilt in the absence of OH. The resulting standing heart rate in POTS patients is generally 120 beats/min [3]. POTS primarily affects women of childbearing age [5,6]. Patients with POTS and OH report debilitating symptoms, including lightheadedness, tachycardia, presyncope, nausea, headache, difficulty concentrating, and memory problems [5].

The relationship between autoimmune disease and orthostatic intolerance is well established. Case reports of patients with autoimmune-related OH have been published. Furthermore, 20% of POTS patients have a diagnosis of a coexisting autoimmune disease, including but not limited to Hashimoto's thyroiditis, celiac disease, Sjogren's disease, rheumatoid arthritis, and systemic lupus erythematosus [5–7].

Interest in underlying autoimmune process in POTS started decades ago. The initial evidence of autoantibodies (AAbs) in POTS patients was reported by Vernino et al. in 2000 when AAbs targeting ganglionic receptors were identified in 7% of POTS patients while not being found in healthy controls [8]. A study by Wallukat et al. suggested an autoimmune mechanism, supported by isolation of AAbs targeting the beta 2-adrenergic receptor (β2AR), the muscarinic M2 receptor (M2R), and the angiotensin II type 1 receptor

(AT1R) [9]. In addition to ganglionic receptor AAbs, identification of beta 1 adrenergic receptor AAbs (β1AR), β2AR AAbs, and muscarinic 3 receptor (M3R) AAbs was reported in 2012 by Yu et al. [10]. More recently, AT1R AAbs were found [11].

The diagnosis of orthostatic intolerance is often preceded by a viral illness or vaccination [12–16]. In the COVID-19 era, orthostatic intolerance was frequently encountered post SARS-CoV-2 infection and vaccination, providing further evidence of an autoimmune etiology [16,17]. In this article, we review the available literature correlating autoimmunity with orthostatic intolerance syndromes. A computerized search in the PubMed, Medline, and Embase databases was performed to retrieve studies with data on orthostatic intolerance and autoantibodies using the search terms orthostatic intolerance, postural orthostatic tachycardia syndrome, orthostatic hypotension, and autoimmunity. Subsequently, a manual search of the reference lists from the retrieved articles was completed to identify additional articles.

2. Pathophysiology

2.1. Pathophysiology of Orthostatic Hypotension

OH is subdivided into neurogenic and non-neurogenic OH. Neurogenic OH is associated with neurodegenerative disorders, such as multiple system atrophy and Parkinson's disease; autoimmune diseases; and neuropathy-associated conditions such as diabetes [12,18,19]. Upon standing, there is a decrease in circulating blood volume of approximately 500 mL to 1000 mL. The decreased blood volume leads to a decrease in preload, stroke volume, and blood pressure. When baroreceptors sense decreased stretch due to decreased intravascular volume, compensatory sympathetic activation increases heart rate and vascular tone to mitigate the effect of the decreased circulating blood volume [20–22]. In neurogenic OH, there is a lack of increase in vascular tone upon standing due to impairment of norepinephrine release [6].

Non-neurogenic OH is due to different mechanisms, including a decrease in circulating blood volume or medication induced by diuretics and vasodilators [19]. Neurogenic and non-neurogenic OH are differentiated by the difference in heart rate from standing and sitting. In neurogenic OH, the change in heart rate from sitting to standing is less than 15 beats per minute [23,24]. When using heart rate to distinguish between the subtypes of OH, it is important to exclude confounding factors, such as bradyarrhythmias, pacemaker dependence, and atrioventricular nodal blocking agents [25].

2.2. Pathophysiology of Postural Orthostatic Tachycardia Syndrome

POTS is theorized to be the culmination of multiple underlying pathological processes, including peripheral neuropathy/denervation, hypovolemia with altered renin–aldosterone levels, and a hyperadrenergic state. The mechanism of peripheral denervation is similar to that of venous pooling with lack of compensatory physiological responses described for OH [12,26].

Regarding hypovolemia, blood volume is reduced in a majority of POTS patients. Reduced stroke volume in the state of hypovolemia is accompanied by compensatory tachycardia to maintain cardiac output [27]. Several studies reported improvement in the severity of POTS symptoms with acute intravascular volume expansion utilizing intravenous saline or desmopressin [28–30]. In addition to compensatory tachycardia, hypovolemia activates the renin–angiotensin–aldosterone system (RAAS), enhancing renal sodium and water retention and subsequent volume expansion. Some POTS patients with the hypovolemic subtype have inappropriately high levels of angiotensin II with low levels of renin and aldosterone [31].

In hyperadrenergic POTS, upon standing for 10 min there is an associated increase in systolic blood pressure of 10 mm Hg and plasma norepinephrine levels of 600 pg/mL [32]. Hyperadrenergic POTS also has associated symptoms of palpitations, tachycardia, and anxiety. These POTS patients are particularly sensitive to any agents that increase adrenergic activity at small doses that have not been shown to induce hemodynamic change in the general population [5,33]. Orthostatic tachycardia without hypotension is key for the diagnosis of POTS [3].

2.2.1. Adrenergic Receptors

Adrenergic receptors are G coupled protein receptors. Adrenergic receptors are further divided into alpha 1, alpha 2, beta 1, beta 2, and beta 3 receptors [34,35]. Alpha 1 adrenergic receptors (α1ARs) exert effects on the blood vessels, increase contractility of the left ventricle, and promote coronary artery vasoconstriction [30]. Table 1 summarizes the adrenergic receptor locations and their physiological effects.

Table 1. Adrenergic receptors and physiological effects.

Receptor	G Protein	Location	Physiological Effects
α1	Gq	Smooth muscles of blood vessels, heart, urinary tract	Arterial and venous constriction, increased ventricular contractility, urinary retention
α2	Gi	Central nervous system, presynaptic sympathetic nerves	Constriction of smooth muscle
β1	Gs	Heart	Increased heart rate, increased ventricular contractility
β 2	Gs	Lung, genitourinary smooth muscle, gastrointestinal tract, platelets	Increased ventricular contractility, increased heart rate Blood vessel dilation Decreased platelet aggregation
β3	Gi Gs	Heart, genitourinary tract, adipose tissue	Decreased cardiac contractility, increase lipolysis

α1AR: alpha 1 adrenergic receptor, α2AR: alpha 2 adrenergic receptor, β1AR: beta 1 adrenergic receptor, β2AR: beta 2 adrenergic receptor, β3AR: beta 3 adrenergic receptor.

The relationship between adrenergic receptor autoantibodies and POTS has been investigated by several studies. The autoantibody-mediated vasodilation mechanism of POTS is apparent when an individual stands. The augmented sympathetic activity leads to orthostatic tachycardia and palpitations. Several studies have demonstrated that AAbs activating β1AR contribute to orthostatic tachycardia, palpitations, and an enhanced adrenergic response [36,37]. Li et al. was able to successfully isolate the human monoclonal antibody that stimulates Beta 2 adrenergic receptors in a patient with orthostatic hypotension and tachyarrhythmias. The effects observed included arteriolar vasodilation, suggesting that the isolated monoclonal antibody has a role in potent vasodilation by altering the normal physiological response to orthostasis [38].

Yu et al. identified autoantibodies against beta adrenergic receptors in five of six patients with OH by use of the enzyme-linked immunosorbent assay (ELISA). These beta adrenergic autoantibodies resulted in activation of protein kinase A, increased contractile activity in cardiac tissue, and altered peripheral vessel contractility demonstrated by a significant dose-dependent vasodilatory effect in animal models [10]. In a clinical study by Li et al., ELISA was used to identify three patients with idiopathic OH and four patients with diabetic OH with more β2AR activation than the healthy controls. The study also demonstrated dose-dependent vasodilation in a rat cremasteric arteriolar assay [39]. In the studies of both Yu et al. and Li et al., the activity of AAbs was blunted by use of propranolol [10,39].

Further studies have demonstrated the presence of AAbs to alpha adrenergic receptors in addition to beta adrenergic receptors. A subsequent study by Li et al. evaluating POTS patients demonstrated the presence of activating β1AR AAbs in all 14 patients and of β2AR autoantibodies in 7 of 14 patients. Evidence of α1AR receptor partial agonist AAbs was also found in POTS patients when their sera infusion caused blunted phenylephrine response in a rat cremaster arteriolar assay [37].

Similar results were found in a study of 17 POTS patients by Fedorowski et al. In this study, 11 of the POTS patients had β1AR AAbs, 12 of the POTS patients had β2AR AAbs, and 8 of the POTS patients had α1AR AAbs [36]. Gunning et al. evaluated 55 POTS patients for adrenergic and muscarinic antibodies. A total of 49 of 55 patients in this study exhibited elevation of α1AR AAbs [40].

Kharrazia et al. measured receptor activity rather than directly measuring AAbs. The receptor activity of all measured receptors—α1AR, β2AR, M2R, and opioid receptor-like 1—was found to be higher in POTS patients compared to controls. Of importance is that the study demonstrated that severity of POTS was correlated strongly with the presence of α1AR [41].

In contrast to most of the studies presented, a study by Hall et al. found no significant difference in 11 antibody levels of adrenergic, muscarinic, angiotensin II, and endothelin between POTS patients and healthy controls [42]. Table 2 summarizes the available evidence for adrenergic antibodies in patients with orthostatic intolerance.

Table 2. Adrenergic antibodies in patients with orthostatic intolerance.

Author	Year	Group of Patients Tested	Receptor-Associated Autoantibodies	Number of Patients with Positive Autoantibodies/Number of Total Patients, Number of Controls with Positive Antibodies/Number of Total Controls	Comments
Yu et al. [10]	2012	Idiopathic OH	β1/2AR	5/6, 0/10	
Li et al. [39]	2012	10 idiopathic OH and 10 diabetic patients with OH	β2AR	7/20, 0/10	
Li et al. [37]	2014	POTS POTS POTS	β1AR β2AR α1AR	14/14, 0/10 7/14, 0/10 14/14, 0/10	
Fedorowski et al. [6]	2016	POTS POTS POTS	α1AR β1AR β2AR	8/17, 0/11 11/17, 0/11 12/17, 0/11	
Gunning et al. [40]	2019	POTS POTS	α1AR	49/55, N/A	
Hall et al. [42]	2022	POTS Control	ELISA: AT1R ETR α1AR α2AR β1AR β2AR M1R M2R M3R M4R M5R	 41/116, 22/81 24/116, 15/81 114/116, 81/81 31/116, 22/81 11/116, 7/81 9/116, 5/81 N/A N/A 24/116, 23/81 28/116, 15/81 N/A	No statistically significant difference in 11 autoantibody levels (adrenergic, muscarinic, angiotensin II, and endothelin) was found between POTS patients and healthy controls

α1AR: alpha 1 adrenergic receptor, AT1R: angiotensin receptor, β1AR: beta 1adrenergic receptor, β2AR: beta 2 adrenergic receptors ETR: endothelin receptor, M1R: muscarinic 1 receptors, M2R: muscarinic 2 receptors, M3R:muscarinic 3 receptors; M4R: muscarinic 4 receptors; M5R muscarinic 5 receptors OH: orthostatic hypotension, POTS: postural orthostatic tachycardia syndrome.

2.2.2. Cholinergic Receptors

Cholinergic receptors are activated by acetylcholine and broadly divided into nicotinic (nAChRs) and muscarinic receptors (mAChRs). The nAChR is found postsynaptically in all autonomic ganglions and at the neuromuscular junction. The mAChR is further categorized into three subtypes, M1, M2, and M3. M1 receptors (M1Rs) are involved in central nervous system transmission. M2 receptors (M2Rs) and M3 receptors (M3Rs) affect exocrine function, gastrointestinal motility, the cardiovascular system, and the airways. Muscarinic receptors are present on the endothelial cells of blood vessels. Although these endothelial muscarinic receptors are not innervated, activation of these receptors by circulating molecules causes vasodilation. In the heart, muscarinic receptors decrease heart rate and slow atrioventricular conduction [2].

In a case study of a patient with OH, the initial presentation was a syncopal event. Subsequent encounters revealed recurrent syncopal events and vital signs consistent with OH. The neurological findings of ptosis and bilateral pupil dilation with diminished pupillary reactivity prompted investigation with a paraneoplastic panel, revealing elevated titers of AChR antibodies. This led to the diagnosis of autoimmune autonomic ganglionopathy (AAG) [6].

In a recent study of 10 POTS patients, 5 patients had elevated M2R AAb levels while none of the controls had elevated M2R AAb activity. The antibody demonstrated a dose-dependent response to increased M2R activation. Furthermore, these antibodies attenuated the response to the M2R agonist oxotremorine. These M2R AAbs may contribute to excessive orthostatic tachycardia due to enhanced withdrawal of vagal tone upon standing [43]. Another study demonstrated a significant association between gastrointestinal symptoms in patients with POTS and levels of mAChR autoantibodies. This is particularly important as it highlights the correlation between the presence of AAbs and the clinical manifestations of POTS [44]. Several studies reported the presence of nicotinic antibodies in patients with orthostatic intolerance. A correlation between the seropositive patients and other dysautonomic manifestations, such as neurogenic bladder and the sicca complex, was demonstrated [8,45–47].

Watarai et al. provided further evidence of the autoimmune basis of POTS by examining the presence of AAbs in POTS patients and patients with neurally mediated syncope. mAChR AAbs were found in 10 of the POTS patients. They occurred with greater frequency in the POTS patients compared to the neurally medicated syncope patients [48].

While autoantibody presence can hint at the autoimmune etiology of orthostatic intolerance, the presence of AAbs is not always clinically significant. Bryarly et al. demonstrated that very low levels and low levels of gACh could be found in the sera of POTS patients and controls. Furthermore, there was no clinical difference between the seropositive POTS patients and the seronegative POTS patients [49].

Table 3 summarizes the studies that investigated anticholinergic antibodies in patients with orthostatic intolerance.

Table 3. Muscarinic receptor antibodies in patients with orthostatic intolerance.

Author	Year	Receptor	Patient Population	Number of Patients with Positive Antibodies/Number of Total Patients, Number of Controls with Positive Antibodies/Number of Total Controls	Comments
Vernino et al. [8] Sandroni et al. [39]	2000 2004	A3-AChR Ab Ganglionic AChR	POTS Orthostatic intolerance, autonomic neuropathy	6/67, N/A Compared 19 seropositive with 87 seronegative patients	Seropositive patients are more likely to have orthostatic hypotension with other cholinergic symptoms like the sicca complex or GI symptoms
Gibbons et al. [50]	2008	AChR	Autoimmune autonomic ganglionopathy	3/3, N/A	Three patients with dysautonomia and nicotine receptor antibody refractory to medical treatment who responded to immunomodulatory therapy

Table 3. *Cont.*

Author	Year	Receptor	Patient Population	Number of Patients with Positive Antibodies/Number of Total Patients, Number of Controls with Positive Antibodies/Number of Total Controls	Comments
McKeon et al. [45]	2009	A3-AChR Ab	Paraneoplastic neurological ganglionopathy	155/15,000 (1%) with positive titers were examined; 13 had pan-dysautonomia, 5 had orthostatic hypotension only	High antibody values of 1.00 nmol/L were associated with pan-dysautonomia
Gibbons et al. [47]	2009	AChR	Autoimmune autonomic ganglionopathy	8/8, N/A	Higher antibody titers were associated with more severe orthostatic hypotension
Yu et al. [10]	2012	M2/M3 receptor Ab	Idiopathic OH	6/6, 0/10	Serum from patients caused dose-dependent vasodilation in rat cremaster arteriole
Li et al. [39]	2012	M3R	10 idiopathic OH and 10 diabetic patients with OH	13/20, 0/10	Serum from patients caused vasodilation in rat cremaster arteriole. The effect was dose-dependent and inhibited by adding atropine
Li et al. [43]	2022	M2R	POTS	5/10, 0/10	These antibodies suppressed the function of M2R in a dose-dependent fashion and may contribute to excessive orthostatic tachycardia due to enhanced withdrawal of vagal tone upon standing
Sunami et al. [44]	2022	mAChR	POTS	N/A	Significant association between gastrointestinal symptoms in patients with POTS disease and level of mAChR autoantibodies
Vernino et al. [8] McKeon et al. [45] Sandroni et al. [46] Gibbons et al. [47]	2000 2009 2004 2009	nAChR	OI		Found a correlation between the seropositive patients and other dysautonomic manifestations, such as neurogenic bladder and the sicca complex
Fedorowski et al. [6]	2022	AChR	OH	N/A, N/A	Case study where OH was recognized as a part of autoimmune autonomic neuropathy
Watari et al. [48]	2018	AChRa3 AChRb4	POTS	8/34, 1/34 2/34, 0/34	
Rodriguez et al. [51]	2021	mAChR	POTS	4/6, N/A	Improvement of symptoms after IVIG treatment

AChR: cholinergic receptor, GI: gastrointestinal, mAChr: muscarinic receptor, M2R:muscarinic 2 receptor, M3R: muscarinic 3 receptor, nAChR: nicotinic receptor, OH: orthostatic hypotension, OI: orthostatic intolerance, POTS: postural orthostatic tachycardia syndrome.

2.2.3. Angiotensin II Type I Receptors

Inappropriately high levels of angiotensin II with low levels of renin and aldosterone in POTS patients have been previously reported. Despite the high levels of angiotensin II, the pressor response is reportedly absent as patients have normal blood pressure. It has been postulated that the blunted pressor response is due to the persistently high levels of angiotensin II as well as reduced activity of angiotensin converting enzyme 2 (ACE2) [31,52,53].

More recently, the role of antibodies against A1TR has emerged. AT1R is another G-coupled protein similar to adrenergic receptors. In a pilot study by Yu et al., serum samples from 17 patients with POTS, 6 patients with recurrent vasovagal syncope (VVS), and 10 controls were obtained. This study demonstrated significant AT1R activity using separated IgG from POTS serum samples as compared to VVS and healthy controls. This AT1R activity was reduced after using losartan, an AT1R blocker. The results of this study provided evidence for the presence of AT1R antibodies in POTS patients [11].

2.2.4. COVID-19 and POTS

The COVID-19 pandemic has generally been associated with respiratory symptoms in the acute phase. However, there have been a variety of symptoms associated with post-acute COVID-19 infection. Long COVID is a term that includes ongoing symptomatic COVID-19 (4 to 12 weeks) and post-COVID-19 syndrome (>12 weeks) that are not explained by an alternative diagnosis [54]. Case reports have described a new onset of autonomic dysfunction symptoms with features of POTS/inappropriate sinus tachycardia in the post-acute phase of COVID-19 infection [55,56]. The underlying cause of dysautonomia post-COVID-19 infection is not well understood. However, a viral infection by SARS-CoV-2 triggering autoimmune response and direct neurotoxic effects has been suggested as an underlying cause for developing post-COVID-19 POTS [57,58]. Wallukat et al. conducted a study in which 31 patients with POTS and COVID-19 were examined. All patients had positive autoantibodies ranging from two to seven. The autoantibodies that were most frequently positive were β2AR, M2R, and AT1R. The presence of β2AR exerted a positive chronotropic effect, the presence of M2AR exerted a negative chronotropic effect, and AT1R exerted a positive chronotropic effect on their targets [9].

A retrospective case series by Blishteyn et al. evaluated patients with no history of chronic orthostatic intolerance. They found evidence of orthostatic intolerance following SARS-CoV-2 infection. In the study there were 15 POTS patients, 2 patients with OH, and 3 patients with neurocardiogenic syncope. Four of the twenty patients had elevated autoimmune/inflammatory markers. Seventeen of the patients had residual autonomic effects that negatively impacted their lives 6 months following SARS-CoV-2 infection. For 12 of the patients, orthostatic intolerance was severe enough to preclude a return to work [16].

Not only does contracting SARS-CoV-2 infection confer the possibility of developing long COVID and subsequent development of POTS, but recipients of the SARS-CoV-2 vaccine have also been shown to develop POTS at a higher rate. The proposed explanation is that an immunological response was elicited by the administration of the vaccination, resulting in similar symptoms to long COVID. The study compared two cohorts, one whose members received the SARS-CoV-2 vaccine and another whose members were positive for SARS-CoV-2 infection to evaluate the diagnosis of POTS both before and after exposure to the vaccine or infection. It was determined that the SARS-CoV-2 vaccine was associated with a statistically significant increase in the development of POTS; however, this increase was less than the development of POTS following SARS-CoV-2 infection [14].

2.2.5. Antiphospholipid and Antinuclear Autoantibodies

Blitshteyn performed a retrospective review of POTS patients to evaluate comorbid autoimmune disorders and the presence of AAbs. Autoimmune disorders were present in 20/100 patients, including Hashimoto's thyroiditis (11/100), antiphospholipid syndrome

(5/100), rheumatoid arthritis (4/100), celiac disease (3/100), systemic lupus erythematosus (2/100), and Sjögren's syndrome (2/100). Antinuclear autoantibodies (ANAs) were positive in 25/100 POTS patients, while antiphospholipid autoantibodies (aPLs) were positive in 7/100 POTS patients. A higher prevalence of ANA AAbs and aPL AAbs were found in patients with POTS compared to the general population. The presence of Hashimoto's thyroiditis, systemic lupus erythematosus, and rheumatoid arthritis was found to be statistically significantly higher in POTS patients compared to the general population [59].

A retrospective case series by Schofield et al. identified 15 patients with antiphospholipid syndrome and orthostatic intolerance. Regarding orthostatic intolerance, 8/15 had POTS, 8/15 had neurocardiogenic syncope, and 3/15 had orthostatic hypotension. Comorbid autoimmune conditions included rheumatoid arthritis (1/15), systemic lupus erythematosus (2/15), and celiac disease (1/15). Two of the POTS patients failed to improve with standard treatment for antiphospholipid syndrome but subsequently responded well to IVIG [60].

2.2.6. Treatment

In addition to orthostatic intolerance being associated with increased mortality, there is a significant impact on quality of life that can prove devastating [12,61]. The initial approach to managing POTS/OH is usually non-pharmacological, including avoidance of triggers (exposure to heat and prolonged standing), graded exercise training, using waist-high compression stockings, and liberal salt and fluid intake [27]. If the non-pharmacological approach is proven to be inadequate, several off-label medications that have demonstrated symptomatic improvement will be administered [12]. These medications include fludrocortisone, ivabradine, beta blockers, midodrine, and pyridostigmine. These medications can be utilized as monotherapy or more often as a combination therapy [12]. For patients with OH, there has been success in treatment with l-threo-3,4-dihidroxyphenylserine (l-DOPS), a synthetic catecholamine that converts to norepinephrine when ingested orally [7,62,63]. Nonetheless, one-third of POTS patients remain symptomatic despite escalation of medical therapy [64]

The question arose whether the available knowledge pertaining to the autoimmunity in orthostatic intolerance patients would predict a role for immune modulation therapeutics in patients with refractory orthostatic intolerance and evidence of existing AAbs. Several case reports and case series demonstrate that immune modulation agents have a possible role in the treatment of orthostatic intolerance in patients who have symptoms refractory to the commonly used pharmacological and non-pharmacological treatments.

Pitarokoili et al. reported a case study in which a female patient with Marfan's syndrome developed POTS 2 weeks following administration of pneumococcal vaccination. Antibodies against adrenergic β1 and β2, muscarinic M2 and M4, and nociceptin-like receptors were positive. She was treated with intravenous immunoglobulin (IVIG), which resulted in improvement of orthostatic symptoms and decreased AAb activity of adrenergic β1 and β2, muscarinic M2 and M4, and nociceptin-like receptors. Maintenance therapy was changed to subcutaneous immunoglobulin. This patient was also able to decrease the dose of clonidine and discontinue midodrine [65]. Another case report for a woman with POTS and mast cell activation syndrome showed improvement in her tachycardia and sudomotor function after 10 IVIG treatments [66].

A large retrospective study by Schofield et al. evaluated the use of IVIG in patients with refractory autoimmune dysautonomias. After being treated with IVIG for at least 3 months, patients experienced improvement in dysautonomic symptoms. The study also demonstrated that the presence of aPL AAbs and Sjögren AAbs correspond to a positive response to IVIG [67]

A case series investigated the role of IVIG treatment in POTS patients. Autoimmune testing revealed that all six patients had AAbs against α1AR, while four of six patients had AAbs against mAChR. After 6 months of IVIG treatment, all patients reported less fatigue, improvement of orthostatic intolerance, and improvement of sudomotor function. Five of six patients described improved exercise tolerance and gastrointestinal symptoms [51].

Kesterson et al. presented a case series of seven patients with POTS treated with subcutaneous immunoglobulin or plasmapheresis. Two patients had positive nAChR at low titers; one patient had elevated adrenergic, muscarinic, AT1R, and endothelin I receptor antibodies; and two patients did not have any identifiable AAbs. The outcomes showed significant improvement in orthostatic symptoms and functional abilities measured by questionnaires preimmunotherapy, 3 months post-treatment, and 12 months post-treatment. Reduction or discontinuation of oral POTS medications was reported among the patients [68].

While IVIG has shown positive response with improvement of symptoms in POTS patients, there are limited data of any benefit on OH patients in the setting of autoimmune autonomic ganglionopathy. A case series explored three patients who did not respond to the conventional methods of fludrocortisone, midodrine, vasopressin, and erythropoetin; plasmapheresis alone; and IVIG alone. They were treated with combination prednisone and mycophenolate mofetil for 6 months followed by five cycles of plasmapheresis. After the course of treatment, OH resolved and mean antibody levels decreased. These results indicate that patients with refractory autoimmune autonomic ganglionopathy may benefit from a multimodal approach to therapy to treat OH [50].

To date, there has been only one randomized control trial, iSTAND, that has evaluated the efficacy and safety of IVIG in POTS patients with moderate to severe symptoms and evidence of autoimmunity either by the presence of AAbs or the coexistence of defined autoimmune diseases. Thirty participants were randomized to receive either IVIG or albumin infusions. COMPASS-31 scores were used to assess symptom response to IVIG and albumin infusions. The iSTAND trial, while not showing a significant difference in symptom outcomes between the IVIG and albumin groups, highlights the challenges in determining optimal treatment strategies for POTS patients with evidence of autoimmunity. The fact that the authors suggested that volume expansion could have been treated with IVIG and albumin emphasizes the complexity of managing these patients and the need for individualized approaches [69]

3. Discussion

This review highlights the contribution of AAbs in symptoms triggered by upright position. AAbs targeting adrenergic receptors cause dose-dependent vasodilation by activating β2AR and partial α1AR antagonism. Impaired vasoconstriction will be paired with tachycardia and palpitations upon standing [10,37,40]. Moreover, β1AR AAbs enhance sympathetic response, causing excessive tachycardia and palpitations with upright position [36,37]. M2R AAbs suppress the function of M2R, with subsequent increased vagal tone withdrawal upon standing [43]. Furthermore, recently discovered AAbs against AT1R may play a role in the state of hypovolemia and RAAS imbalance in a subgroup of POTS patients [11]. In a study by Gunning et al., the detection of α1AR-AAbs in POTS patients was coupled with significant elevation of several cytokines compared to control subjects, shedding light on autoimmunity in POTS and the autoinflammatory state in this disease [70].

3.1. *Importance of a Reliable Methodology*

In order to be able to effectively target the patients with an autoimmune component of POTS, AAbs need to be properly identified. The studies presented in this review use various methods for the identification of AAbs. Hall et al. provided evidence that there was no significant difference between controls and POTS patients by using ELISA [42]. This finding is contrary to what other studies investigating orthostatic intolerance have determined. It was postulated by the authors that this discrepancy was due to theirs being the first study to evaluate POTS with a control group. Several of the studies completed prior to Hall et al. did have a control group for orthostatic hypotension and demonstrated significant differences in AAbs between patients with OH and the controls [10,39]. However, it should be noted that the size of the studies was smaller, which may affect the reliability of the studies. Future studies should include a larger quantity of participants to determine if ELISA can reliably be used as a methodology to identify AAbs. Furthermore, it may be of interest to identify AAbs with ELISA and a different assay to compare detection of AAbs with different assays in syndromes of orthostatic intolerance.

3.2. Individualizing Treatment Plans

The variety of treatment options for orthostatic intolerance serve as a reminder that treatment is not straightforward; an individualized approach is needed. This is particularly evident with the use of IVIG, as the iSTAND trial demonstrated no significant difference in symptoms when compared to the control group [51] In order to formulate an individualized approach, the exact role of AAbs in orthostatic intolerance needs to be further elucidated. Much of the available literature reports the effects of AAb stimulation on receptors; however, little is known about the mechanism by which AAbs exert their effects.

While there is a dearth of studies explaining the mechanism of AAbs in orthostatic intolerance, there are a few studies that provide valuable insights. Deng et al. investigated M2R-AAbs in a rabbit model. Treatment of M2R-immunized rabbits with low-level tragus stimulation (LLTS) was performed to stimulate the vagus nerve. LLTS treatment resulted in blunting of postural tachycardia, increased acetylcholine secretion, and improved the attenuated chronotropic heart rate response. This study provides insight into the mechanism of M2R AAbs, showing that by increasing the production of acetylcholine, the effects of M2R AAbs can be overcome. Further insight into the mechanism is provided with the noted decrease in inflammatory cytokines following treatment with LLTS [71] Guo et al. performed a similar study with alpha adrenergic receptors and beta adrenergic receptors in a rabbit model. The results indicated that there was increased release of acetylcholine and elevated inflammatory markers [72].

Stavrakis et al. provided evidence of the mechanism of AAb-mediated POTS by evaluating transcutaneous vagal stimulation in a randomized control trial. The results were decreases in β1AR and α1AR autoantibodies, improvement in cardiac autonomic function, and a decrease in serum inflammatory cytokines. From a POTS symptom standpoint, patients experienced less sudomotor stimulation and decreased orthostatic tachycardia [73]. Of particular importance is the noted decrease in adrenergic AAb production. These studies illuminate the mechanism by which AAbs exert their effects and demonstrate a cost-effective method for treatment of autoimmune-mediated POTS that can be incorporated into an individualized treatment plan. Future studies are needed to understand how increased parasympathetic stimulation leads to decreased AAb production.

Once identified, it is important to clinically correlate symptoms with AAbs. Bryarly et al. demonstrated that the mere presence of very low and low levels of an antibody does not indicate clinical significance [49]. This is important, as further costs to the patient can be avoided if the clinician recognizes that there is no need for further immune workup with low titers. This has implications for treatment, as in such a case the patient would not be classified according to the autoimmune etiology of orthostatic intolerance and other treatment options could be explored.

A universal definition of what constitutes the autoimmune etiology of POTS by titer levels would be useful. The application of a universally agreed upon definition would directly impact studies. The iSTAND trial evaluated patients with suspected autoimmune etiology of POTS; however, the study failed to mention what the titer levels were for patients. The patients who received IVIG in the study could have had low titer levels, resulting in no difference to the control group [69].

As previously mentioned, there are several postulated mechanisms for POTS. POTS patients have reported a delay in diagnosis, likely due to the complexities associated with understanding the diagnosis of POTS [5]. While the focus of this review is on studies related to autoimmune-mediated orthostatic intolerance, the presence of AAbs is not necessary for the diagnosis of orthostatic intolerance. If general practitioners are cognizant of the association of orthostatic intolerance with autoimmune etiology, this may prompt them to consider the diagnosis of POTS or obtain an early referral to a specialist, such as a neurologist or a cardiologist, who can confirm the diagnosis and provide further management. With evidence that many patients suffering from long COVID can develop autoimmune-mediated orthostatic intolerance, prompt recognition is important as it can expedite appropriate treatment [9]. By recognizing the autoimmune component of both OH and POTS, an investigation can be quickly started to identify any possible associated autoimmune disorder and expedite early treatment with immunomodulating therapies.

4. Conclusions

The review provides a comprehensive overview of the existing evidence linking orthostatic intolerance to autoimmunity. Many studies have shown increased levels of AAbs in patients with orthostatic intolerance when compared to controls. The only randomized control study evaluating IVIG for autoimmunity failed to show the benefit of IVIG over standard methods of volume expansion. This highlights the complexity of managing orthostatic intolerance and emphasizes the need for larger randomized control trials to explore the specificity of AAbs. While the effects of AAbs on receptors are known, further studies evaluating the mechanism could aid in the development of more treatment options.

Author Contributions: Conceptualization: B.G.; Methodology: A.M., Writing: C.P., A.M. and A.-R.M.; Review and editing: B.G. and A.M.; Supervision: B.G. All authors have read and agreed to the published version of the manuscript.

Funding: This research received no external funding.

Institutional Review Board Statement: Not applicable.

Informed Consent Statement: Not applicable.

Data Availability Statement: No new data were created or analyzed in this study. Data sharing is not applicable to this article.

Conflicts of Interest: The authors declare no conflicts of interest.

References

1. Stewart, J.M. Common syndromes of orthostatic intolerance. *Pediatrics* **2013**, *131*, 968–980. [CrossRef] [PubMed]
2. Ruzieh, M.; Batizy, L.; Dasa, O.; Oostra, C.; Grubb, B. The role of autoantibodies in the syndromes of orthostatic intolerance: A systematic review. *Scand. Cardiovasc. J.* **2017**, *51*, 243–247. [CrossRef] [PubMed]
3. Freeman, R.; Wieling, W.; Axelrod, F.B.; Benditt, D.G.; Benarroch, E.; Biaggioni, I.; Cheshire, W.P.; Chelimsky, T.; Cortelli, P.; Gibbons, C.H.; et al. Consensus statement on the definition of orthostatic hypotension, neurally mediated syncope and the postural tachycardia syndrome. *Clin. Auton. Res.* **2011**, *21*, 69–72. [CrossRef] [PubMed]
4. Fedorowski, A.; Stavenow, L.; Hedblad, B.; Berglund, G.; Nilsson, P.M.; Melander, O. Orthostatic hypotension predicts all-cause mortality and coronary events in middle-aged individuals (The Malmo Preventive Project). *Eur. Heart J.* **2010**, *31*, 85–91. [CrossRef] [PubMed]
5. Shaw, B.H.; Tiles, L.E.; Bourne, K.; Green, E.A.; Shibao, C.A.; Okamoto, L.E.; Garland, E.M.; Gamboa, A.; Diedrich, A.; Raj, V.; et al. The face of postural tachycardia syndrome—Insights from a large cross-sectional online community-based survey. *J. Intern. Med.* **2019**, *286*, 438–448. [CrossRef] [PubMed]
6. Fedorowski, A.; Ricci, F.; Hamrefors, V.; Sandau, K.E.; Hwan Chung, T.; Muldowney, J.A.S.; Gopinathannair, R.; Olshansky, B. Orthostatic Hypotension: Management of a Complex, But Common, Medical Problem. *Circ. Arrhythmia Electrophysiol.* **2022**, *15*, e010573, Erratum in *Circ. Arrhythmia Electrophysiol.* **2023**, *16*, 478. [CrossRef] [PubMed]
7. Gibbons, C.H.; Verino, S.A.; Kaufmann, H.; Freeman, R. L-DOPS therapy for refractory orthostatic hypotension in autoimmune autonomic neuropathy. *Neurology* **2005**, *65*, 1104–1106. [CrossRef] [PubMed]
8. Vernino, S.; Low, P.A.; Fealey, R.D.; Stewart, J.D.; Farrugia, G.; Lennon, V.A. Autoantibodies to ganglionic acetylcholine receptors in autoimmune autonomic neuropathies. *N. Engl. J. Med.* **2000**, *343*, 847–855. [CrossRef]
9. Wallukat, G.; Hohberger, B.; Wenzel, K.; Fürst, J.; Schulze-Rothe, S.; Wallukat, A.; Hönicke, A.S.; Müller, J. Functional autoantibodies against G-protein coupled receptors in patients with persistent Long-COVID-19 symptoms. *J. Transl. Autoimmun.* **2021**, *4*, 100100. [CrossRef]

10. Yu, X.; Stavrakis, S.; Hill, M.A.; Huang, S.; Reim, S.; Li, H.; Khan, M.; Hamlett, S.; Cunningham, M.W.; Kem, D.C. Autoantibody activation of beta-adrenergic and muscarinic receptors contributes to an "autoimmune" orthostatic hypotension. *J. Am. Soc. Hypertens* **2012**, *6*, 40–47. [CrossRef]
11. Yu, X.; Li, H.; Murphy, T.A.; Yu, X.; Li, H.; Murphy, T.A.; Nuss, Z.; Liles, J.; Liles, C.; Aston, C.E.; et al. Angiotensin II Type 1 Receptor Autoantibodies in Postural Tachycardia Syndrome. *J. Am. Heart Assoc.* **2018**, *7*, e008351. [CrossRef] [PubMed]
12. Sheldon, R.S.; Grubb, B.P., 2nd; Olshansky, B.; Shen, W.K.; Calkins, H.; Brignole, M.; Raj, S.R.; Krahn, A.D.; Morillo, C.A.; Stewart, J.M.; et al. 2015 heart rhythm society expert consensus statement on the diagnosis and treatment of postural tachycardia syndrome, inappropriate sinus tachycardia, and vasovagal syncope. *Heart Rhythm.* **2015**, *12*, e41–e63. [CrossRef]
13. Garland, E.M.; Celedonio, J.E.; Raj, S.R. Postural Tachycardia Syndrome: Beyond Orthostatic Intolerance. *Curr. Neurol. Neurosci. Rep.* **2015**, *15*, 60. [CrossRef] [PubMed]
14. Kwan, A.C.; Ebinger, J.E.; Wei, J.; Le, C.N.; Oft, J.R.; Zabner, R.; Teodorescu, D.; Botting, P.G.; Navarrette, J.; Ouyang, D.; et al. Apparent Risks of Postural Orthostatic Tachycardia Syndrome Diagnoses After COVID-19 Vaccination and SARS-CoV-2 Infection. *Nat. Cardiovasc. Res.* **2022**, *1*, 1187–1194. [CrossRef] [PubMed]
15. Blitshteyn, S. Postural tachycardia syndrome following human papillomavirus vaccination. *Eur. J. Neurol.* **2014**, *21*, 135–139. [CrossRef] [PubMed]
16. Blitshteyn, S.; Whitelaw, S. Postural orthostatic tachycardia syndrome (POTS) and other autonomic disorders after COVID-19 infection: A case series of 20 patients. *Immunol. Res.* **2021**, *69*, 205–211. [CrossRef] [PubMed]
17. Ormiston, C.K.; Świątkiewicz, I.; Taub, P.R. Postural orthostatic tachycardia syndrome as a sequela of COVID-19. *Heart Rhythm.* **2022**, *19*, 1880–1889. [CrossRef] [PubMed]
18. Isaacson, S.H.; Dashtipour, K.; Mehdirad, A.A.; Peltier, A.C. Management Strategies for Comorbid Supine Hypertension in Patients with Neurogenic Orthostatic Hypotension. *Curr. Neurol. Neurosci. Rep.* **2021**, *21*, 18. [CrossRef] [PubMed]
19. Fedorowski, A.; Melander, O. Syndromes of orthostatic intolerance: A hidden danger. *J. Intern. Med.* **2013**, *273*, 322–335. [CrossRef]
20. Ricci, F.; De Caterina, R.; Fedorowski, A. Orthostatic Hypotension: Epidemiology, Prognosis, and Treatment. *J. Am. Coll. Cardiol.* **2015**, *66*, 848–860. [CrossRef]
21. Lanier, J.B.; Mote, M.B.; Clay, E.C. Evaluation and management of orthostatic hypotension. *Am. Fam. Physician* **2011**, *84*, 527–536.
22. Kaufmann, H.; Norcliffe-Kaufmann, L.; Palma, J.-A. Baroreflex dysfunction. *N. Engl. J. Med.* **2020**, *382*, 163–178. [CrossRef]
23. Freeman, R.; Abuzinadah, A.R.; Gibbons, C.; Jones, P.; Miglis, M.G.; Sinn, D.I. Orthostatic hypotension: JACC state-of-the-art review. *J. Am. Coll. Cardiol.* **2018**, *72*, 1294–1309. [CrossRef]
24. Olshansky, B.; Muldowney, J. Cardiovascular safety considerations in the treatment of neurogenic orthostatic hypotension. *Am. J. Cardiol.* **2020**, *125*, 1582–1593. [CrossRef]
25. Gibbons, C.H.; Schmidt, P.; Biaggioni, I.; Frazier-Mills, C.; Freeman, R.; Isaacson, S.; Karabin, B.; Kuritzky, L.; Lew, M.; Low, P.; et al. The recommendations of a consensus panel for the screening, diagnosis, and treatment of neurogenic orthostatic hypotension and associated supine hypertension. *J. Neurol.* **2017**, *264*, 1567–1582. [CrossRef]
26. Van Lieshout, J.J.; Wieling, W.; Karemaker, J.M.; Secher, N.H. Syncope, cerebral perfusion, and oxygenation. *J. Appl. Physiol.* **2003**, *94*, 833–848. [CrossRef]
27. Vernino, S.; Bourne, K.M.; Stiles, L.E.; Grubb, B.P.; Fedorowski, A.; Stewart, J.M.; Arnold, A.C.; Pace, L.A.; Axelsson, J.; Boris, J.R.; et al. Postural orthostatic tachycardia syndrome (POTS): State of the science and clinical care from a 2019 National Institutes of Health Expert Consensus Meeting—Part 1. *Auton. Neurosci.* **2021**, *235*, 102828. [CrossRef] [PubMed]
28. Coffin, S.T.; Black, B.K.; Biaggioni, I.; Paranjape, S.Y.; Orozco, C.; Black, P.W.; Dupont, W.D.; Robertson, D.; Raj, S.R. Desmopressin acutely decreases tachycardia and improves symptoms in the postural tachycardia syndrome. *Heart Rhythm.* **2012**, *9*, 1484–1490. [CrossRef]
29. Fu, Q.; Vangundy, T.B.; Galbreath, M.M.; Shibata, S.; Jain, M.; Hastings, J.L.; Bhella, P.S.; Levine, B.D. Cardiac origins of the postural orthostatic tachycardia syndrome. *J. Am. Coll. Cardiol.* **2010**, *55*, 2858–2868. [CrossRef] [PubMed]
30. Jacob, G.; Shannon, J.R.; Black, B.; Biaggioni, I.; Mosqueda-Garcia, R.; Robertson, R.M.; Robertson, D. Effects of volume loading and pressor agents in idiopathic orthostatic tachycardia. *Circulation* **1997**, *96*, 575–580. [CrossRef]
31. Raj, S.R.; Biaggioni, I.; Yamhure, P.C.; Black, B.K.; Paranjape, S.Y.; Byrne, D.W.; Robertson, D. Renin-aldosterone paradox and perturbed blood volume regulation underlying postural tachycardia syndrome. *Circulation* **2005**, *111*, 1574–1582. [CrossRef] [PubMed]
32. Garland, E.M.; Raj, S.R.; Black, B.K.; Harris, P.A.; Robertson, D. The hemodynamic and neurohumoral phenotype of postural tachycardia syndrome. *Neurology* **2007**, *69*, 790–798. [CrossRef] [PubMed]
33. Abe, H.; Nagatomo, T.; Kohshi, K.; Numata, T.; Kikuchi, K.; Sonoda, S.; Mizuki, T.; Kuroiwa, A.; Nakashima, Y. Heart rate and plasma cyclic AMP responses to isoproterenol infusion and effect of beta-adrenergic blockade in patients with postural orthostatic tachycardia syndrome. *J. Cardiovasc. Pharmacol.* **2000**, *36* (Suppl. 2), S79–S82. [CrossRef] [PubMed]
34. Motiejunaite, J.; Amar, L.; Vidal-Petiot, E. Adrenergic receptors and cardiovascular effects of catecholamines. *Ann. Endocrinol.* **2021**, *82*, 193–197. [CrossRef] [PubMed]
35. Schena, G.; Caplan, M.J. Everything You Always Wanted to Know about β3-AR * (But Were Afraid to Ask). *Cells* **2019**, *8*, 357. [CrossRef] [PubMed]

36. Fedorowski, A.; Li, H.; Yu, X.; Koelsch, K.A.; Harris, V.M.; Liles, C.; Murphy, T.A.; Quadri, S.M.S.; Scofield, R.H.; Sutton, R.; et al. Antiadrenergic autoimmunity in postural tachycardia syndrome. *EP Europace* **2017**, *19*, 1211–1219. [CrossRef] [PubMed]
37. Li, H.; Yu, X.; Liles, C.; Khan, M.; Vanderlinde-Wood, M.; Galloway, A.; Zillner, C.; Benbrook, A.; Reim, S.; Collier, D.; et al. Autoimmune basis for postural tachycardia syndrome. *J. Am. Heart Assoc.* **2014**, *3*, e000755. [CrossRef] [PubMed]
38. Li, H.; Zuccolo, J.; Kem, D.C.; Zillner, C.; Lee, J.; Smith, K.; James, J.A.; Cunningham, M.W.; Yu, X. Implications of a vasodilatory human monoclonal autoantibody in postural hypotension. *J. Biol. Chem.* **2013**, *288*, 30734–30741. [CrossRef] [PubMed]
39. Li, H.; Kem, D.C.; Reim, S.; Khan, M.; Vanderlinde-Wood, M.; Zillner, C.; Collier, D.; Liles, C.; Hill, M.A.; Cunningham, M.W.; et al. Agonistic autoantibodies as vasodilators in orthostatic hypotension: A new mechanism. *Hypertension* **2012**, *59*, 402–408. [CrossRef] [PubMed]
40. Gunning, W.T., 3rd; Kvale, H.; Kramer, P.M.; Karabin, B.L.; Grubb, B.P. Postural Orthostatic Tachycardia Syndrome Is Associated with Elevated G-Protein Coupled Receptor Autoantibodies. *J. Am. Heart Assoc.* **2019**, *8*, e013602. [CrossRef]
41. Kharraziha, I.; Axelsson, J.; Ricci, F.; Di Martino, G.; Persson, M.; Sutton, R.; Fedorowski, A.; Hamrefors, V. Serum Activity Against G Protein-Coupled Receptors and Severity of Orthostatic Symptoms in Postural Orthostatic Tachycardia Syndrome. *J. Am. Heart Assoc.* **2020**, *9*, e015989. [CrossRef]
42. Hall, J.; Bourne, K.M.; Vernino, S.; Hamrefors, V.; Kharraziha, I.; Nilsson, J.; Sheldon, R.S.; Fedorowski, A.; Raj, S.R. Detection of G Protein-Coupled Receptor Autoantibodies in Postural Orthostatic Tachycardia Syndrome Using Standard Methodology. *Circulation* **2022**, *146*, 613–622. [CrossRef]
43. Li, H.; Zhang, G.; Forsythe, E.; Okamoto, L.E.; Yu, X. Implications of Antimuscarinic Autoantibodies in Postural Tachycardia Syndrome. *J. Cardiovasc. Transl. Res.* **2022**, *15*, 438–440. [CrossRef]
44. Sunami, Y.; Sugaya, K.; Miyakoshi, N.; Iwazaki, O.; Takahashi, K. Association of autoantibodies to muscarinic acetylcholine receptors with gastrointestinal symptoms and disease severity in patients with postural orthostatic tachycardia syndrome. *Immunol. Res.* **2022**, *70*, 197–207. [CrossRef]
45. McKeon, A.; Lennon, V.A.; Lachance, D.H.; Fealey, R.D.; Pittock, S.J. Ganglionic acetylcholine receptor autoantibody: Oncological, neurological, and serological accompaniments. *Arch. Neurol.* **2009**, *66*, 735–741. [CrossRef] [PubMed]
46. Sandroni, P.; Vernino, S.; Klein, C.M.; Lennon, V.A.; Benrud-Larson, L.; Sletten, D.; Low, P.A. Idiopathic autonomic neuropathy: Comparison of cases seropositive and seronegative for ganglionic acetylcholine receptor antibody. *Arch. Neurol.* **2004**, *61*, 44–48. [CrossRef]
47. Gibbons, C.H.; Freeman, R. Antibody titers predict clinical features of autoimmune autonomic ganglionopathy. *Auton. Neurosci.* **2009**, *146*, 8–12. [CrossRef]
48. Watari, M.; Nakane, S.; Mukaino, A.; Nakajima, M.; Mori, Y.; Maeda, Y.; Masuda, T.; Takamatsu, K.; Kouzaki, Y.; Higuchi, O.; et al. Autoimmune postural orthostatic tachycardia syndrome. *Ann. Clin. Transl. Neurol.* **2018**, *5*, 486–492. [CrossRef] [PubMed]
49. Bryarly, M.; Raj, S.R.; Phillips, L.; Hynan, L.S.; Okamoto, L.E.; Arnold, A.C.; Paranjape, S.Y.; Vernino, M.; Black, B.K.; Vernino, S. Ganglionic Acetylcholine Receptor Antibodies in Postural Tachycardia Syndrome. *Neurol. Clin. Pract.* **2021**, *11*, e397–e401. [CrossRef]
50. Gibbons, C.H.; Vernino, S.A.; Freeman, R. Combined Immunomodulatory Therapy in Autoimmune Autonomic Ganglionopathy. *Arch Neurol.* **2008**, *65*, 213–217. [CrossRef] [PubMed]
51. Rodriguez, B.; Hoepner, R.; Salmen, A.; Kamber, N.; Z'Graggen, W.J. Immunomodulatory treatment in postural tachycardia syndrome: A case series. *Eur. J. Neurol.* **2021**, *28*, 1692–1697. [CrossRef] [PubMed]
52. Mustafa, H.I.; Raj, S.R.; Diedrich, A.; Black, B.K.; Paranjape, S.Y.; Dupont, W.D.; Williams, G.H.; Biaggioni, I.; Robertson, D. Altered systemic hemodynamic and baroreflex response to angiotensin II in postural tachycardia syndrome. *Circ. Arrhythmia Electrophysiol.* **2012**, *5*, 173–180. [CrossRef] [PubMed]
53. Stewart, J.M.; Ocon, A.J.; Clarke, D.; Taneja, I.; Medow, M.S. Defects in cutaneous angiotensin-converting enzyme 2 and angiotensin-(1-7) production in postural tachycardia syndrome. *Hypertension* **2009**, *53*, 767–774. [CrossRef] [PubMed]
54. NICE. *COVID-19 Rapid Guideline: Managing the Long-Term Effects of COVID-19*; NICE: London, UK, 2020.
55. Desai, A.D.; Boursiquot, B.C.; Moore, C.J.; Gopinathannair, R.; Waase, M.P.; Rubin, G.A.; Wan, E.Y. Autonomic dysfunction post–acute COVID-19 infection. *HeartRhythm Case Rep.* **2022**, *8*, 143–146. [CrossRef] [PubMed]
56. Gupta, A.; Madhavan, M.V.; Sehgal, K.; Nair, N.; Mahajan, S.; Sehrawat, T.S.; Bikdeli, B.; Ahluwalia, N.; Ausiello, J.C.; Wan, E.Y.; et al. Extrapulmonary manifestations of COVID-19. *Nat. Med.* **2020**, *26*, 1017–1032. [CrossRef] [PubMed]
57. Mallick, D.; Goyal, L.; Chourasia, P.; Zapata, M.R.; Yashi, K.; Surani, S. COVID-19 Induced Postural Orthostatic Tachycardia Syndrome (POTS): A Review. *Cureus* **2023**, *15*, e36955. [CrossRef] [PubMed]
58. El-Rhermoul, F.-Z.; Fedorowski, A.; Eardley, P.; Taraborrelli, P.; Panagopoulos, D.; Sutton, R.; Lim, P.B.; Dani, M. Autoimmunity in Long Covid and POTS. *Oxf. Open Immunol.* **2023**, *4*, iqad002. [CrossRef] [PubMed]
59. Blitshteyn, S. Autoimmune markers and autoimmune disorders in patients with postural tachycardia syndrome (POTS). *Lupus* **2015**, *24*, 1364–1369. [CrossRef] [PubMed]
60. Schofield, J.R.; Blitshteyn, S.; Shoenfeld, Y.; Hughes, G.R. Postural tachycardia syndrome (POTS) and other autonomic disorders in antiphospholipid (Hughes) syndrome (APS). *Lupus* **2014**, *23*, 697–702. [CrossRef] [PubMed]
61. Masaki, K.H.; Schatz, I.J.; Burchfiel, C.M.; Sharp, D.S.; Chiu, D.; Foley, D.; Curb, J.D. Orthostatic hypotension predicts mortality in elderly men: The Honolulu Heart Program. *Circulation* **1998**, *98*, 2290–2295. [CrossRef]

62. Goldstein, D.S. L-Dihydroxyphenylserine (L-DOPS): A Norepinephrine Prodrug. *Cardiovasc. Drug Rev.* **2006**, *24*, 189–203. [CrossRef] [PubMed]
63. Sánchez-Ferro, A.; Benito-León, J.; Gómez-Esteban, J.C. The management of orthostatic hypotension in Parkinson's disease. *Front. Neurol.* **2013**, *4*, 64. [CrossRef]
64. Arnold, A.C.; Ng, J.; Raj, S.R. Postural tachycardia syndrome–Diagnosis, physiology, and prognosis. *Auton Neurosci.* **2018**, *215*, 3–11. [CrossRef]
65. Pitarokoili, K.; Maier, A.; de Moya Rubio, E.C.; Hahn, K.; Wallukat, G.; Athanasopoulos, D.; Grüter, T.; Motte, J.; Fisse, A.L.; Gold, R. Maintenance therapy with subcutaneous immunoglobulin in a patient with immune-mediated neuropathic postural tachycardia syndrome. *J. Transl. Autoimmun.* **2021**, *4*, 100112. [CrossRef]
66. Weinstock, L.B.; Brook, J.B.; Myers, T.L.; Goodman, B. Successful treatment of postural orthostatic tachycardia and mast cell activation syndromes using naltrexone, immunoglobulin and antibiotic treatment. *BMJ Case Rep.* **2018**, *2018*, bcr2017221405. [CrossRef] [PubMed]
67. Schofield, J.R.; Chemali, K.R. Intravenous Immunoglobulin Therapy in Refractory Autoimmune Dysautonomias: A Retrospective Analysis of 38 Patients. *Am. J. Ther.* **2019**, *26*, 570–582. [CrossRef] [PubMed]
68. Kesterson, K.; Schofield, J.; Blitshteyn, S. Immunotherapy with subcutaneous immunoglobulin or plasmapheresis in patients with postural orthostatic tachycardia syndrome (POTS). *J. Neurol.* **2023**, *270*, 233–239. [CrossRef] [PubMed]
69. Vernino, S.; Hopkins, S.; Bryarly, M.; Hernandez, R.S.; Salter, A. Randomized controlled trial of intravenous immunoglobulin for autoimmune postural orthostatic tachycardia syndrome (iSTAND). *Clin. Auton. Res.* **2024**, *34*, 153–163. [CrossRef]
70. Gunning, W.T.; Kramer, P.M.; Cichocki, J.A.; Karabin, B.L.; Khuder, S.A.; Grubb, B.P. Platelet Storage Pool Deficiency and Elevated Inflammatory Biomarkers Are Prevalent in Postural Orthostatic Tachycardia Syndrome. *Cells* **2022**, *11*, 774. [CrossRef]
71. Deng, J.; Li, H.; Guo, Y.; Zhang, G.; Fischer, H.; Stavrakis, S.; Yu, X. Transcutaneous vagus nerve stimulation attenuates autoantibody-mediated cardiovagal dysfunction and inflammation in a rabbit model of postural tachycardia syndrome. *J. Interv. Card Electrophysiol.* **2023**, *66*, 291–300. [CrossRef]
72. Guo, Y.; Li, H.; Deng, J.; Zhang, G.; Fischer, H.; Stavrakis, S.; Yu, X. Low-level tragus stimulation improves autoantibody-induced hyperadrenergic postural tachycardia syndrome in rabbits. *Heart Rhythm O2* **2022**, *4*, 127–133. [CrossRef] [PubMed]
73. Stavrakis, S.; Chakraborty, P.; Farhat, K.; Whyte, S.; Morris, L.; Abideen Asad, Z.U.; Karfonta, B.; Anjum, J.; Matlock, H.G.; Cai, X.; et al. Noninvasive Vagus Nerve Stimulation in Postural Tachycardia Syndrome: A Randomized Clinical Trial. *JACC Clin. Electrophysiol.* **2024**, *10*, 346–355. [CrossRef] [PubMed]

Journal of *Personalized Medicine*

MDPI

Communication

The Head-Up Tilt Table Test as a Measure of Autonomic Functioning among Patients with Myalgic Encephalomyelitis/Chronic Fatigue Syndrome

Leonard A. Jason [1,*], William J. McGarrigle [2] and Ruud C. W. Vermeulen [3]

1 Department of Psychology, DePaul University, Chicago, IL 60614, USA
2 Department of Psychology, University of Kentucky, Lexington, KY 40508, USA; wmcgarrigle@uky.edu
3 CFS Medical Center, 1114AA Amsterdam, The Netherlands; rv@cvsmemc.nl
* Correspondence: ljason@depaul.edu

Abstract: Individuals with Myalgic Encephalomyelitis/Chronic Fatigue Syndrome (ME/CFS) often experience autonomic symptoms. In the present study, we evaluated 193 adults seeking treatment for ME/CFS, who were recruited from an outpatient clinic. The participants completed a head-up tilt table test to assess two common types of orthostatic intolerance, namely, postural orthostatic tachycardia syndrome (POTS) and orthostatic hypotension (OH). During the tilt test, 32.5% of the participants demonstrated POTS or OH. The participants with either of these two common types of orthostatic intolerance were found to have more problems with sleep and post-exertional malaise as assessed by the DePaul Symptom Questionnaire; these patients also reported more physical and health function limitations. The implications of the findings are discussed.

Keywords: Myalgic Encephalomyelitis/Chronic Fatigue Syndrome; tilt table test; orthostatic intolerance

1. Introduction

Patients with Myalgic Encephalomyelitis/Chronic Fatigue Syndrome (ME/CFS) often experience autonomic symptoms, including nausea, headaches, sleep disturbance, and cognitive problems [1]. Patients with ME/CFS also commonly demonstrate elevations in their resting heart rate, systolic blood pressure, and mean arterial blood pressure, and often show a lower stroke index [2]. Up to 75% of adults with ME/CFS have these symptoms, which could be due to a malfunctioning autonomic nervous system [3].

One type of autonomic dysfunction implicated in ME/CFS is orthostatic intolerance (OI), signifying abnormal dynamic blood pressure regulation [4]. OI is defined by an inability to tolerate an upright position and is relieved by rest and recumbence [5]. Common symptoms of OI include dizziness, lightheadedness, and syncope, among others. One type of OI in patients with ME/CFS is vasovagal syncope (also called simple fainting or neurocardiogenic or neurally mediated syncope). Two other common subtypes are postural orthostatic tachycardia syndrome (POTS) and orthostatic hypotension (OH). POTS is marked by a substantial increase in the heart rate when transitioning from the supine to an upright position [6], whereas OH involves a fall in blood pressure upon standing [7]. These conditions contribute to ME/CFS symptoms broadly and are associated with a decrease in the quality of life [8]. Schultz, Katz, Bockian, and Jason [9] found significant correlations between youth self-reported levels of orthostatic and autonomic functioning and physician measurement of orthostatic functioning; however, that study did not involve a head-up tilt table test, which serves as an autonomic assessment of OI.

A head-up tilt table test can assess these two common types of OI—POTS and OH. The head-up tilt table test is one of the major assessment tools that has been used in ME/CFS research trials. Another stressor assessment tool that has been used to assess physiological systems involving ME/CFS is a two-day cardiopulmonary exercise test (CPET) performed

24 h apart [10]. In patients with ME/CFS, the CPET demonstrates an inability to reproduce maximal or anaerobic threshold measures on the second day; values on the second CPET are much lower than those on the first CPET. However, this test may induce severe exacerbation of symptoms in these patients. Due to this, several researchers [11] have suggested using a single-day CPET, but a single day does not capture the aerobic impairment. Keller, Pryor, and Giloteaux [11] found that a single CPET resulted in the classification of 12 of 22 patients with ME/CFS as having little or no impairment, and 8 patients as having mild/moderate impairment. But data from the second day's CPET indicated that aerobic energy-producing processes failed to respond normally to exercise. As Batemen et al. [10] suggest, the CPET should be used for disability testing only, as these tests involve a stressor that may induce severe or long-lasting post-exertional malaise.

The present study involved a sample of ME/CFS patients from the Netherlands. We explored the percentage of patients who had OI (using either tilt-table testing or self-reports) so that we could determine how common OI is among patients with ME/CFS. We were interested in exploring whether OI has a high prevalence among patients with ME/CFS, such as post-exertional malaise, cognitive impairment, and unrefreshing sleep. If patients with ME/CFS are selected from tertiary care settings that specialize in OI, it is more likely that OI would be a prevalent symptom of ME/CFS, but they might occur less frequently in non-tertiary care settings. In other words, differences could be due to where the patients are recruited, as specialty clinics tend to attract more severely impaired patients [3]. The present study hypothesized that patients recruited from a setting that did not specialize in OI care might have lower rates of OI. The following study used the head-up tilt table test to assess POTS and OH among an adult sample of patients with ME/CFS, and the outcomes of this test were related to self-reported symptoms and overall functioning.

2. Method

2.1. Participants

The sample consisted of 193 adults with a physician report of ME/CFS and referred to an outpatient clinic in the Netherlands (the CFS Medical Center in Amsterdam).

2.2. Tilt Table Test Procedure

The head-up tilt table test [12] served as an autonomic assessment of orthostatic intolerance. During the test, which lasted 20 min, an appropriately sized cuff was placed on the participant's upper arm and the participant was instructed to remain still and silent in the supine position for 10 min. At that time, blood pressure and pulse were measured with a sphygmomanometer (Omron M6). After 10 min, the table was raised to a 70-degree head-up tilt for another 10 min. The test was terminated after 10 min in the tilted position or sooner if the participant reported complaints indicating insufficient cerebral perfusion.

The participants were labeled as positive for orthostatic intolerance (OI+) if they demonstrated either POTS or OH during the head-up tilt table test. OI− indicated that the patient did not have POTS or OH. POTS was defined as a heart rate increase of $\geq$30 bpm that is sustained (i.e., lasting at least two consecutive minutes) within 10 min of the tilt; OH was defined as a sustained decrease of at least 20 mmHg in systolic blood pressure or 10 mm Hg of diastolic blood pressure within 3 min of the tilt [13]. Baseline blood pressure and pulse were defined by those collected at the ninth minute of the test, just before the tilt.

2.3. Measures

2.3.1. The DePaul Symptom Questionnaire

The DePaul Symptom Questionnaire (DSQ-1) was completed by the participants. The DSQ-1 is a 54-item self-report that measures ME/CFS symptomology, demographics, and medical, occupational, and social history [14]. Using a 5-point Likert scale, the DSQ-1 indexes the frequency and severity of symptoms within the past 6 months. The scale for frequency is as follows: 0 = *none of the time*, 1 = *a little of the time*, 2 = *about half of the time*, 3 = *most of the time*, and 4 = *all of the time*. The scale for severity is as follows: 0 = symptom not present, 1 = *mild*, 2 = *moderate*, 3 = *severe*, 4 = *very severe*. The scores are converted into a 100-point scale, and the frequency and severity scores of each symptom are averaged into one composite score of the symptom.

The DSQ-1 has shown good to excellent test–retest reliability for those with ME/CFS and individuals within the control groups [15]. Factor analytic studies using this instrument have identified cardinal symptom clusters, or core domains, of ME/CFS [16]. The DSQ-1 has been used to differentiate ME/CFS from other chronic illnesses, like multiple sclerosis [17]. The Shared Library of Research Electronic Data Capture (REDCap) offers access to the DSQ-1 through the host of DePaul University.

2.3.2. The 36-Item Short-Form Health Survey (SF-36)

The participants in this study also completed the 36-item Short-Form Health Survey (SF-36). The SF-36 is a self-report inventory that focuses on eight different domains: Physical Functioning, Role Physical, Bodily Pain, General Health Perceptions, Vitality, Social Functioning, Role Emotional, and Mental Health [18]. Items are rated on a five-point Likert scale, with higher scores indicating better health, or that a patient's functioning is being less impacted by their health. The SF-36 is considered a reliable and valid instrument capable of differentiating between patient and non-patient populations [19].

3. Results

3.1. Demographics

The age ranged from 18 to 68 years ($M = 38.3$, $SD = 12.16$). Most participants were female (78.8%). A duration of illness longer than two years was reported by 67.9% of the sample. Regarding work status, patients were categorized as either Working Part- or Full-Time/Students (52.8%) or Disabled/Unemployed/Retired (47.2%). There were no significant differences observed between the OI+ and OI− groups on demographic characteristics.

3.2. Outcomes

During the tilt table test, 32.5% ($n = 63$) of the participants demonstrated orthostatic intolerance (POTS or OH). Table 1 provides the differences between the OI+ and OI− groups for the main DSQ-1 domains and symptoms. On average, OI− group had significantly lower scores (i.e., less frequent and severe) on the Sleep and Post-Exertional Malaise symptom domain and the following symptom items: unrefreshing sleep, difficulty falling asleep, difficulty staying asleep, waking up early, trouble forming words, and feeling chills or shivers. As displayed in Table 2, the OI+ group had significantly greater impairment in the SF-36 Physical Functioning and General Health domains.

Table 1. Significant Domain and Symptom Differences.

	OI− (n = 130) *M* (SD)	OI+ (n = 63) *M* (SD)	*p*
Sleep Domain	50.83 (17.24)	43.71 (17.95)	0.01
Unrefreshed	86.29 (14.16)	79.75 (12.24)	0.01
Difficulty falling asleep	55.11 (29.39)	42.36 (27.55)	0.00
Difficulty staying asleep	53.50 (32.05)	43.06 (30.97)	0.02
Waking up early	40.46 (32.99)	32.29 (30.72)	0.07
PEM Domain	72.90 (17.19)	64.79 (19.26)	0.00
Trouble forming words	58.20 (23.78)	48.50 (22.13)	0.00
Feeling chills or shivers	31.85 (24.42)	21.53 (22.21)	0.00

Table 2. Comparison of SF-36 domain composite scores.

	OI− (*n* = 130) *M* (SD)	OI+ (*n* = 63) *M* (SD)	*p*
Physical Functioning	43.40 (23.46)	36.11 (23.95)	0.04
Role Physical	5.81 (19.64)	3.57 (14.80)	0.42
Bodily Pain	44.81 (24.88)	41.11 (26.40)	0.34
General Health	28.53 (18.08)	20.06 (14.80)	0.00
Vitality	23.92 (15.36)	19.76 (13.18)	0.07
Role Emotional	68.22 (42.63)	66.67 (42.75)	0.81
Mental Health	61.88 (16.65)	60.00 (17.88)	0.48

4. Discussion

This study's major finding is that, on average, patients with POTS or OH experienced more symptoms and functional limitations than those not experiencing POTS or OH. Interestingly, only 32.5% of the participants demonstrated OI (POTS or OH) during the tilt table test. Although it is not surprising that those with POTS or OH have more physical and health functional problems, we expected to find a higher percentage of individuals with POTS or OH. It is possible that the low rate of 32.5% demonstrating POTS or OH could be due to not including other forms of OI, such as vasovagal syncope. It is also possible that cerebral blood flow is reduced in ME/CFS during head-up tilt testing even in the absence of hypotension or tachycardia [20]. It is also plausible that given the high percentage of patients with POTS or OH that were either working or in the student status, this sample had relatively less impairment. If this is the case, we speculate that the head-up tilt test may be more effective at detecting OI among more severely impaired patient cohorts, such as those within tertiary care settings.

In addition to functional limitations, we found that patients with POTS or OH reported elevated scores on the DSQ-1 post-exertional malaise and neurocognitive symptom domains; this group also demonstrated higher composite scores on individual DSQ-1 items, including unrefreshing sleep, difficulty falling asleep, difficulty staying asleep, waking up early, trouble forming words, and feeling chills or shivers. The wide assortment of sleep, neurocognitive, and neuroendocrine features suggests that those with POTS or OH have symptoms that are likely contributing to their functional limitations.

Notably, among our OI+ group, only 37.7% indicated "Feeling unsteady on your feet, like you might fall", using the threshold score of at least moderate severity and at least half the time. Comparable results were found for the item "Dizziness or fainting", where only 32.3% of those with a positive tilt test met the threshold burden of at least moderate severity and frequency of half the time or more. These findings suggest that the majority of individuals in our sample with positive tilt table test data did not meet the threshold for these OI self-report items being a burden to the patients.

The findings from this study have relevance to the Institute of Medicine [21] report that provided a new case definition for ME/CFS. In brief, the new clinical case definition required a substantial reduction in pre-illness levels of activity, post-exertional malaise, unrefreshing sleep, and either neurocognitive impairment and/or OI. The IOM report also operationalized OI as having a moderate or greater frequency and severity of symptoms. These new criteria had some similarities with prior ME/CFS case definitions and their stipulation of symptoms [22,23], but the IOM criteria was the first time an ME/CFS case definition required either neurocognitive impairment and/or OI [24].

Focusing on the IOM case definition, Jason et al. [24] found that 67% of patients with ME/CFS report OI, whereas 93% report cognitive impairment. These researchers found that by using the OI symptoms instead of neurocognitive impairment, only 2% more participants met the IOM criteria than if the criteria had only required cognitive impairment. A different approach was tried by Chu et al. [25], but her team utilized a categorical response of "yes" and "no" to measure "feeling sick, uncomfortable, or fainting while standing." In contrast to Jason et al.'s [24] findings, Chu et al. (2017) found that 92% of participants reported OI and 87% of participants endorsed cognitive impairment. Chu et al.'s group operationalization of OI allowed for 13% more participants to meet the IOM criteria than if participants were required to endorse cognitive impairment alone. Chu and colleagues hypothesized the discrepancy in findings from Jason et al. [24] might have been due to the researchers' use of "less common symptoms" to represent OI (e.g., shortness of breath and irregular heartbeats). Additionally, Chu and colleagues did not require minimum frequency and severity thresholds as required by the IOM.

To deal with this controversy, Gaglio et al. [26] assessed different methods of operationalizing OI for the IOM criteria. With a sample of two-hundred and forty-two participants who completed the DSQ, they examined how many participants met the IOM criteria while endorsing different frequencies and severities of various OI symptoms. While neurocognitive impairment occurred in 93.4% of patients with ME/CFS, OI without concurrent neurocognitive symptoms only allowed for an additional 1.7–4.5% of participants to meet the IOM criteria. These results do not support the IOM's inclusion of neurocognitive impairment and OI as interchangeable symptoms.

Although as indicated in the introduction, OI can result in significant impairment, it has not been found to be among the most prevalent ME/CFS symptoms [24]. Other researchers have found similar results, such as Schondorf et al.'s [27] study where only 40% of their ME/CFS sample had a positive tilt test (indicative of OI). In addition, Timmers et al. [28] reported an even lower percentage of 27.8%. In addition, LaManca et al. [29] were not able to find any significant differences in presyncope symptoms or heart rate and blood pressure changes (indicative of OI assessment) between those with ME/CFS and controls. These studies along with the present study indicate that OI might not be considered a core symptom of ME/CFS.

Still, OI is a symptom of at least some patients with ME/CFS. In those patients with OI, there appears to be either too little or too much expression of insufficient control of the autonomic systems. In the upright position, the pressure in the circulation in the lower part of the body increases and the response is an increase in the tension of the vessel walls. Too little and the blood pools in the lower part and too much increases the resistance, expressed as an increase in the diastolic blood pressure, a lower pulse pressure, and a lower stroke volume. The increase in the heart rate is an attempt to compensate for the lower cardiac output. In ME/CFS, there is also a complicating low blood volume, sometimes comparable to a hypovolemic shock in which lifting of the head results in a major increase in the heart rate. There probably is some interference and bias between symptoms and results of the table test.

There are many other potential biological ways to identify the multiple causes of OI symptoms, including anemia (which can be determined by routine blood tests, as low normal hemoglobin may affect oxygen supply to the brain) [30], oxygen dissociation (in a person with normal hemoglobin and hematocrit, the red blood cells may have a strong affinity for holding onto the oxygen) [31], Ehlers–Danlos syndrome (where the lax blood vessels in lower extremities allow blood to pool and blood does not reach the heart and brain adequately upon standing) [32], vasopressin/ADH deficiency (diabetes insipidus) [33], and low blood volume (which could be related to aldosterone levels) [34] (Jacob et al., 1998). Ryabkova et al. [35] found similar patterns of dysautonomia involving heart rate variability, blood pressure variability, and baroreflex failure in patients with ME/CFS and Long COVID. After the head-up tilt test, Swai et al. [36] found that patients with POTS have lower heart rate variability in terms of time domain measure but not in terms of frequency domain measure. In addition, a subgroup of ME patients have autoantibodies to adrenergic receptors in the central nervous system [37] and this is probably related to OI. Certainly, OI is complex and multiple methods might need to be employed to adequately assess and understand these symptoms.

Physiological testing such as tilt table testing and exercise testing (including VO2 max) have been used to address specific questions, often in consultation with a specialist [38]. Tilting and exercise have been used as a provocation in ME/CFS specifically because they provoke the disease symptoms, thus making it easier to see abnormalities in metabolism, skeletal muscle, gene expression, neurological and cognitive measures, cardiovascular/autonomic reflex abnormalities, immune abnormalities, and oxidative stress and alterations in the microbiome. Keller et al. [11] demonstrate that patients with ME/CFS have a different response to CPET testing and the present study suggests that at least some patients with ME/CFS exhibit OI, but not at the same percentage as other classic ME/CFS symptoms such as post-exertional malaise and cognitive impairment.

This study had several limitations. For example, we did not assess OI symptoms following the tilt table test; follow-up data might have allowed us to better evaluate the effects of this stressor on the patients. In addition, the sample size for the OI+ group was considerably smaller than that of the OI− group. Future studies would benefit from more extensive monitoring of autonomic symptom indicators using larger cohorts of patients with and without OI.

In conclusion, our study found evidence that those with POTS or OH have more limitations as well as symptoms than those without POTS or OH. Even so, only about one-third of the patients had POTS or OH based on the tilt table test. Further research is needed to determine the relationship between positive tilt test data and self-reported symptoms of OI, given that the Institute of Medicine [21] currently lists OI and/or cognitive impairment as one of the defining symptoms of ME/CFS and the present study suggests that OI might be an important feature of ME/CFS but not a core symptom that is essential to the syndrome (i.e., post-exertional malaise, unrefreshing sleep, and cognitive impairment).

Author Contributions: Conceptualization, L.A.J., W.J.M. and R.C.W.V.; methodology, R.C.W.V.; formal analysis, L.A.J. and W.J.M.; resources, L.A.J., W.J.M. and R.C.W.V.; writing—original draft preparation, L.A.J. and W.J.M.; writing—review and editing, L.A.J.; supervision, L.A.J. All authors have read and agreed to the published version of the manuscript.

Funding: This research received no external funding.

Institutional Review Board Statement: Regarding reference number W24.015, the Medical Research Involving Human Subjects Act (WMO) does not apply to this study and therefore an official approval of this study by the Medical Research Ethics Committees United is not required under the WMO.

Informed Consent Statement: Written informed consent was obtained from all subjects involved in the study.

Data Availability Statement: Data will be made available by contacting the first author.

Conflicts of Interest: The authors declare no conflicts of interest.

References

1. Katz, B.; Jason, L. Chronic fatigue following infections in adolescents. *Infect. Dis. Immun.* **2013**, *25*, 95–102. [CrossRef]
2. Newton, J.; Okonkwo, O.; Sutcliffe, K.; Seth, A.; Shin, J.; Jones, D. Symptoms of autonomic dysfunction in chronic fatigue syndrome. *QJM* **2007**, *100*, 519–526. [CrossRef]
3. Van Cauwenbergh, D.; Nijs, J.; Kos, D.; Van Weijnen, L.; Struyf, F.; Meeus, M. Malfunctioning of the autonomic nervous system in patients with chronic fatigue syndrome: A systematic literature review. *Eur. J. Clin. Investig.* **2014**, *44*, 516–526. [CrossRef]
4. Novak, P. Cerebral blood flow, heart rate, and blood pressure patterns during the tilt test in common orthostatic syndromes. *Neurosci. J.* **2016**, *2016*, 6127340. [CrossRef] [PubMed]
5. Lambert, E.; Lambert, G.W. Sympathetic Dysfunction in Vasovagal Syncope and the Postural Orthostatic Tachycardia Syndrome. *Front. Physiol.* **2014**, *5*, 280. [CrossRef] [PubMed]
6. Lewis, I.; Pairman, J.; Spickett, G.; Newton, J.L. Clinical characteristics of a novel subgroup of chronic fatigue syndrome patients with postural orthostatic tachycardia syndrome. *J. Intern. Med.* **2013**, *273*, 501–510. [CrossRef] [PubMed]
7. Fedorowski, A.; Melander, O. Syndromes of orthostatic intolerance: A hidden danger. *J. Intern. Med.* **2012**, *273*, 322–335. [CrossRef]
8. Stewart, J.M. Update on the theory and management of orthostatic intolerance and related syndromes in adolescents and children. Expert Rev. *Cardiovasc. Ther.* **2012**, *10*, 1386–1399. [CrossRef]

9. Schultz, K.R.; Katz, B.Z.; Bockian, N.R.; Jason, L.A. Associations between autonomic and orthostatic self-report and physician ratings of orthostatic intolerance in youth. *Clin. Ther.* **2019**, *41*, 633–640. [CrossRef]
10. Bateman, L.; Bested, A.C.; Bonilla, H.F.; Chheda, B.V.; Chu, L.; Curtin, J.M.; Dempsey, T.T.; Dimmock, M.E.; Dowell, T.G.; Felsenstein, D.; et al. Myalgic Encephalomyelitis/Chronic Fatigue Syndrome: Essentials of Diagnosis and Management. *Mayo Clin. Proc.* **2021**, *96*, 2861–2878. [CrossRef]
11. Keller, B.A.; Pryor, J.L.; Giloteaux, L. Inability of myalgic encephalomyelitis/chronic fatigue syndrome patients to reproduce VO_2peak indicates functional impairment. *J. Transl. Med.* **2014**, *12*, 104. [CrossRef]
12. Kenny, R.A.; Bayliss, J.; Ingram, A.; Sutton, R. Head-up tilt: A useful test for investigating unexplained syncope. *Lancet* **1986**, *1*, 1352–1355. [CrossRef]
13. Freeman, R.; Wieling, W.; Axelrod, F.B.; Benditt, D.G.; Benarroch, E.; Biaggioni, I.; Cheshire, W.P.; Chelimsky, T.; Cortelli, P.; Gibbons, C.H.; et al. Consensus statement on thedefinition of orthostatic hypotension, neurally mediated syncope andthe postural tachycardia syndrome. *Auton. Neurosci.* **2011**, *21*, 69–72.
14. Jason, L.A.; Sunnquist, M. The development of the DePaul Symptom Questionnaire: Original, expanded, brief, and pediatric versions. *Front. Pediatr.* **2018**, *6*, 330. [CrossRef]
15. Jason, L.A.; So, S.; Brown, A.A.; Sunnquist, M.; Evans, M. Test-retest reliability of the DePaul Symptom Questionnaire. *Fatigue Biomed. Health Behav.* **2015**, *3*, 16–32. [CrossRef]
16. Conroy, K.E.; Islam, M.F.; Jason, L.A. Evaluating Case Diagnostic Criteria for Myalgic Encephalomyelitis/Chronic Fatigue Syndrome (ME/CFS): Toward an empirical case definition. *Disabil. Rehabil.* **2023**, *45*, 840–847. [CrossRef]
17. Ohanian, D.; Brown, A.; Sunnquist, M.; Furst, J.; Nicholson, L.; Klebek, L.; Jason, L.A. Identifying key symptoms differentiating Myalgic Encephalomyelitis and Chronic Fatigue Syndrome from Multiple Sclerosis. *EC Neurol.* **2016**, *4*, 41–45.
18. Ware, J.E., Jr.; Sherbourne, C.D. The MOS 36-item short-form health survey (SF-36): I. Conceptual framework and item selection. *Medcare* **1992**, *30*, 473–483. [CrossRef]
19. McHorney, C.A.; Ware, J.E., Jr.; Lu, J.F.R.; Sherbourne, C.D. The MOS 36-ltem short-form Health survey (SF-36): III. Tests of data quality, scaling assumptions, and reliability across diverse patient groups. *Medcare* **1994**, *32*, 40–66. [CrossRef] [PubMed]
20. van Campen, C.M.; Verheugt, F.W.A.; Rowe, P.C.; Visser, F.C. Cerebral blood flow is reduced in ME/CFS during head-up tilt testing even in the absence of hypotension or tachycardia: A quantitative, controlled study using Doppler echography. *Clin. Neurophysiol. Pract.* **2020**, *5*, 50–58. [CrossRef] [PubMed]
21. Institute of Medicine. *Beyond Myalgic Encephalomyelitis/Chronic Fatigue Syndrome: Redefining an Illness*; The National Academies Press: Washington, DC, USA, 2015. [CrossRef]
22. Carruthers, B.M.; Jain, A.K.; De Meirleir, K.L.; Peterson, D.L.; Klimas, N.G.; Lerner, A.M.; Bested, A.C.; Flor-Henry, P.; Joshi, P.; Powles, A.C.P.; et al. Myalgic Encephalomyelitis/Chronic Fatigue Syndrome: Clinical Working Case Definition, Diagnostic and Treatment Protocols. *J. Chronic Fatigue Syndr.* **2003**, *11*, 7–115. [CrossRef]
23. Fukuda, K.; Straus, S.E.; Hickie, I.; Sharpe, M.C.; Dobbins, J.G.; Komaroff, A.; Group ICFSS. The Chronic Fatigue Syndrome: A Comprehensive Approach to its Definition and Study. *Ann. Intern. Med.* **1994**, *121*, 953–959. [CrossRef]
24. Jason, L.A.; Sunnquist Brown, M.; Newton, J.L.; Strand, E.B.; Vernon, S.D. Chronic Fatigue Syndrome versus Systemic Exertion Intolerance Disease. *Fatigue Biomed. Health Behav.* **2015**, *3*, 127–141. [CrossRef]
25. Chu, L.; Norris, J.L.; Valencia, I.J.; Montoya, J.G. Patients Diagnosed with Myalgic Encephalomyelitis/Chronic Fatigue Syndrome Also Fit Systemic Exertion Intolerance Disease Criteria. *Fatigue: Biomed. Health Behav.* **2017**, *5*, 114–128. [CrossRef]
26. Gaglio, C.L.; Islam, M.F.; Cotler, J.; Jason, L.A. Orthostatic Intolerance and neurocognitive impairment in Myalgic Encephalomyelitis/Chronic Fatigue Syndrome (ME/CFS). *Epidemiol. Methods* **2022**, *11*, 20210033. [CrossRef]
27. Schondorf, R.; Benoit, J.; Wein, T.; Phaneuf, D. Orthostatic Intolerance in the Chronic Fatigue Syndrome. *J. Auton. Nerv. Syst.* **1999**, *75*, 192–201. [CrossRef]
28. Timmers, H.J.; Wieling, W.; Soetekouw, P.M.; Bleijenberg, G.; Van Der Meer, J.W.; Lenders, J.W. Hemodynamic and Neurohumoral Responses to Head-Up Tilt in Patients with Chronic Fatigue Syndrome. *Clin. Auton. Res.* **2002**, *12*, 273–280. [CrossRef]
29. LaManca, L.; Peckerman, A.; Walker, J.; Kesil, W.; Cook, S.; Taylor, A.; Natelson, B.H. Cardiovascular Response during Head-Up Tilt in Chronic Fatigue Syndrome. *Clin. Physiol. Funct. Imaging* **1999**, *19*, 111–120. [CrossRef]
30. Perera, R.; Isola, L.; Kaufmann, H. Effect of recombinant erythropoietin on anemia and orthostatic hypotension in primary autonomic failure. *Clin. Auton. Res.* **1995**, *5*, 211–213. [CrossRef]
31. Kharraziha, I.; Holm, H.; Bachus, E.; Ricci, F.; Sutton, R.; Fedorowski, A.; Hamrefors, V. Cerebral Oximetry in Syncope and Syndromes of Orthostatic Intolerance. *Front. Cardiovasc. Med.* **2019**, *6*, 171. [CrossRef]
32. Peebles, K.C.; Tan, I.; Butlin, M.; Collins, F.; Tofts, L.; Avolio, A.P.; Pacey, V. The prevalence and impact of orthostatic intolerance in young women across the hypermobility spectrum. *Am. J. Med Genet. A* **2022**, *188*, 1761–1776. [CrossRef]
33. Zerbe, R.L.; Henry, D.P.; Robertson, G.L. Vasopressin response to orthostatic hypotension: Etiologic and clinical implications. *Am. J. Med.* **1983**, *74*, 265–271. [CrossRef]
34. Jacob, G.; Biaggioni, I.; Mosqueda-Garcia, R.; Robertson, R.M.; Robertson, D. Relation of blood volume and blood pressure in orthostatic intolerance. *Am. J. Med. Sci.* **1998**, *315*, 95–100. [CrossRef]
35. Ryabkova, V.A.; Rubinskiy, A.V.; Marchenko, V.N.; Trofimov, V.I.; Churilov, L.P. Similar patterns of Dysautonomia in Myalgic Encephalomyelitis/Chronic Fatigue and Post-COVID-19 Syndromes. *Pathophysiology* **2024**, *31*, 1–17. [CrossRef]

36. Swai, J.; Hu, Z.; Zhao, X.; Rugambwa, T.; Ming, G. Heart rate and heart rate variability comparison between postural orthostatic tachycardia syndrome versus healthy participants; a systematic review and meta-analysis. *BMC Cardiovasc. Disord.* **2019**, *19*, 320. [CrossRef]
37. Bynke, A.; Julin, P.; Gottfries, C.G.; Heidecke, H.; Scheibenbogen, C.; Bergquist, J. Autoantibodies to beta-adrenergic and muscarinic cholinergic receptors in Myalgic Encephalomyelitis (ME) patients—A validation study in plasma and cerebrospinal fluid from two Swedish cohorts. *Brain Behav. Immun. Health* **2020**, *7*, 100107. [CrossRef]
38. CDC Myalgic Encephalomyelitis/Chronic Fatigue Syndrome. 2023. Available online: https://www.cdc.gov/me-cfs/healthcare-providers/diagnosis/evaluation.html (accessed on 15 December 2023).

Journal of
*Personalized
Medicine*

MDPI

Article

At-Home Blood Pressure Measurements Provide Better Assessments of Orthostatic Hypotension in Parkinson's Disease

Chathurini V Fernando [1,†], Sarah Osborn [2,†] and Malcolm Horne [2,3,*]

1 Florey Institute of Neuroscience and Mental Health, Parkville, VIC 3010, Australia; chathurini.fernando@florey.edu.au
2 The Bionics Institute, East Melbourne, VIC 3002, Australia; sosborn@bionicsinstitute.org
3 Department of Medicine, University of Melbourne, St Vincent's Hospital, Fitzroy, VIC 3065, Australia
* Correspondence: mhorne@unimelb.edu.au
† These authors contributed equally to this work.

Abstract: Orthostatic hypotension (OH) is common in Parkinson's Disease (PD). It is intermittent, exacerbated by stressors including meals, medications, and dehydration, and frequently is unrecognized. Although intermittent, assessment is usually by a single "in clinic" BP measurement. This study examines whether 10 home measurements are more sensitive in detecting OH than a single "in clinic" measurement. Participants (44 people with PD and 16 controls) were instructed to measure lying and standing BP at home. BP was measured on five consecutive days upon waking and before bedtime. Symptoms were also assessed using the Movement Disorder Society United Parkinson's Disease Rating Scale and the Non-Motor Questionnaire. While a postural drop in systolic BP ($\geq$20 mmHg) was recorded "in clinic" in thirteen of the forty-four PD participants, a postural drop was found in at least one of the ten home measurements in twenty-eight of the forty-four participants. Morning hypertension and variability in lying systolic BP was more common in these subjects than in those without a postural drop or the controls. A greater number of measurements of lying and standing BP are more likely to reveal orthostatic hypotension, variation in systolic BP, and hypertension than a single office measurement in people with PD.

Keywords: Parkinson's Disease; orthostatic hypotension; hypertension; cardiovascular dysregulation; autonomic dysfunction

1. Introduction

Orthostatic hypotension (OH) is common in Parkinson's Disease (PD), with a prevalence of between 30% and 50% [1–3]. OH is important because it leads to impaired cerebral perfusion [4], resulting in well-known symptoms [5] of light headedness, dizziness [6], loss of consciousness, and falls [7–9], and has been linked to impaired cognition [10–17] and mortality [18,19].

The differences between the pathophysiology of OH in PD and many other causes of OH can be understood by first reviewing the normal physiological response to an orthostatic challenge. Transferring from lying to standing shifts ~700 mls from the central compartment to lower extremities (~500 mL) and pelvic regions (~200 mL) [20–22] decreases central venous pressure, which is sensed by cardiopulmonary baroreceptors, resulting in reduced baroreflex signaling in the brain stem, which decreases vagal nerve activity and increases sympathetic activity and the release of noradrenaline. This, in turn, increases peripheral resistance, heart rate, and contractility [20–22]. Thus, there are both central and peripheral mechanisms of regulation: "central mechanisms" refer collectively to the brainstem and cortical structures that regulate autonomic function and include the dorsal motor nuclei of the vagus, the medullary reticular formation, the locus coeruleus [21,23] and insular cortex [24], and "peripheral mechanisms", referring collectively to vagal and the pre- and postganglionic sympathetic control of end organs. It is important to note for later discussion that central mechanisms selectively control the perfusion of specific vascular beds, depending on their physiological demands. When central pressures are low, this same mechanism prioritises perfusion of the brain, heart, and kidneys over perfusion of other vascular beds. For example, food ingestion is followed by splanchnic vasodilation and the pooling of splanchnic blood, which activates the baroreflex mechanism to maintain normal BP [25]. If, however, this increased cardiac response is insufficient to adequately perfuse the brain and heart (perhaps, for example, because of coexistent hypovolaemia), then splanchnic vasoconstriction would occur, allowing blood volume to be maintained in essential compartments at the expense of the gut.

In the general population, common causes of OH include hypovolemia, polypharmacy, heart failure, arrhythmias, and advanced valvular heart disease [5]. In these conditions, both central and peripheral mechanisms are intact, in contrast with neurogenic OH, which is characterised by the pathological impairment of peripheral autonomic mechanisms. Neurogenic OH occurs in people with spinal cord injuries [26] and small fibre neuropathies, including diabetes [17]. While OH in PD is considered neurogenic in origin [5], it differs from other neurogenic OH because its pathophysiology is contributed to by the impairment of both peripheral and central mechanisms [20]. The baroreflex gain is low [27], indicating a dysfunctional central mechanism. PD pathology is present in the brain stem sites mediating the baroreflex [21,23] and also in the insular cortex [24] (see Ref. [20] for a review). While central control of sympathetic function is disturbed relatively early in PD [28,29], baroreflex failure alone does not usually cause OH [27], as peripheral mechanisms must also be present. Evidence for impaired peripheral mechanisms in OH of PD includes a low noradrenergic response to orthostatic challenge [30–32] and cardiac sympathetic denervation and dysfunction [33] (see [20,34] for a review).

The consequence of this broader autonomic dysregulation in PD is that the combined effect of otherwise minor stressors, such as the vasodilating effect of levodopa, hypertensive agents, exercise, dehydration, and food [1,25,35], cannot be defended against. For example, consider a person who has breakfast in the morning when their BP is already low because of relative dehydration and levodopa-induced vasodilation. Impaired central mechanisms mean that post-breakfast splanchnic vasodilation cannot be inhibited and instead persists, further compounding low BP. Furthermore, cerebral perfusion may be further compromised because cerebrovascular autoregulation is also disturbed [4,36]. Thus, in PD, OH appears intermittently and often in response to a confluence of stressors. On the other hand, supine hypertension may occur [10] because the baroreflex and renal mechanisms are not centrally coordinated to respond to fluid from the lower extremities returning to central compartments overnight. There is also marked variability in systolic BP [37–39], which is frequently elevated in the morning. Capturing these features requires frequent BP measures with morning measurements or measurements when at least one stressor, for example, standing, is present.

In the routine clinical care of PD, OH is usually identified by a single lying and standing systolic BP in the clinic. Performed properly, this requires the person with PD (PwP) to lie resting for 5 min followed by BP measured supine, immediately when standing, and then 3 min later. This is a serious impost on time in a busy practice, and an anecdotal poll of colleagues in private practice suggests that compromises are made and corners are cut, even to the extent of measuring sitting rather than lying BP. Thus, an effective alternative would be welcomed in routine care. Measurement is often prompted by a history of symptoms consistent with OH. However, history is unreliable, with episodes of OH frequently asymptomatic or unrecognised by the PwP [4,37,40], as well as the presence of symptoms not correlating with the presence of OH [37]. Furthermore, as discussed above, OH can be intermittent and thus missed by a single measurement, which also cannot identify variability or morning supine hypertension. Twenty-four-hour blood pressure recordings are frequently used but they do not readily identify stressor-induced drops in BP and, as noted above, OH recognition is low in PD, so self-reporting diaries can fail. A novel pilot study of eight subjects undergoing continuous 5-day monitoring [37] provided results indicating that unrecognised events and systolic variability could be detected by prolonged recordings. Usual OH assessments are lab-based, expensive, and do not address the issues of OH in PD: particularly BP variability and supine hypertension.

This study was directed at the question of whether more frequent measures of lying and standing blood pressure performed at home by the PwP, including early morning measurement, might improve the detection of OH, supine hypertension, and systolic BP variability. PwPs were provided with a calibrated electric sphygmomanometer and were instructed in taking and recording lying and standing BP. They then took twice daily measurements on five consecutive days in their own home. The results were compared to lying and standing BP measured in the clinic.

2. Materials and Methods

This study was approved and overseen by the St. Vincent's Hospital (Melbourne) Human Research and Ethics Committee (approval number LRR 320.21). Subjects provided written consent according to the Declaration of Helsinki, and the study was conducted according to the International Conference on Harmonisation: Good Clinical Practice Guidelines (ICH-GCP).

2.1. Subjects and Recruitment

Participants were 44 PwPs with a history of idiopathic PD and 16 people without PD (controls: usually the spouse of the PwP). All participants were aged 60 years or more. PwPs were required to be 6 or more years from onset of symptoms or diagnosis to increase the likelihood that a significant proportion would have clinical OH at the time of enrollment (27% had postural drop in the clinic plus symptoms, as shown in Table 1) and that a similar proportion would not have OH, even on repeated measures. Cases with other potential causes of OH including a prescription for diuretics, diabetes (requiring insulin), small fibre neuropathy, heart failure, renal failure, or other reasons for fluid volume disturbance were excluded. Medications that could contribute to OH were recorded but, except for diuretics and insulin, were not a cause for exclusion. Antihypertensives were taken by 27% of the PwPs and 47% of the controls. Medications for urinary urgency were taken by 10% of the PwPs and 6% of the controls. Antidepressants were taken by 4% of the PwPs and 6% of the controls. Fludrocortisone was taken by 6% of the PwPs. PwPs were recruited by reviewing the clinic appointment diary to identify subjects who were due to attend the clinic and contacting them by phone to assess their eligibility and willingness to participate.

Table 1. Participants' demographics, BP, and data from clinical scales.

Parameter	Control	PwD
Age	69 (9)	72 (8)
MoCA	26 (3)	24 (5)
Systolic BP	128 (22)	131 (25)
Diastolic BP	74 (12)	77 (14)
Disease Duration		10 (6)
UPDRS I		11 (7)
UPDRS II		15 (12)
UPDRS III		40 (20)
UPDRS IV		6 (6)
UPDRS Total		60 (29)
MDS_H&Y		2 (1)
OHSA TOTAL		0 (3)
OHDAS TOTAL		0 (0)
PDQ 39		21 (44.5)
NMS TOTAL		12 (10)
Prior Diagnosis of OH	1/16 (6%)	12/44 (27%)

All values are the median with the interquartile range (IQR) in brackets. Abbreviations for the clinical scales are defined in Section 2.2.

On the day of attendance at the clinic, written consent to participate was obtained. Participants were provided with instructions for recording lying and standing blood pressure at home (see below). Lying and standing blood pressure was also measured. Clinical scales were administered (see next section). Participants' demographics, medications, and data from various clinical scales were recorded and are shown in Table 1.

2.2. Clinical Scales

Clinical scales performed included the Movement Disorder Society United Parkinson's Disease Rating Scale (MDS-UPDRS), the Montreal Cognitive Assessment (MoCA), the Non-Motor Questionnaire [41] (NMS-Q), the Parkinson's Disease Questionnaire (PDQ-39), and the Orthostatic Hypotension Questionnaire (OHQ).

2.3. Blood Pressure Recordings

Participants were provided with an Omron HEM 7121 electronic BP machine that was calibrated by the hospital's biomedical engineering department. Instructions for recording BP were:

- Attach the cuff to the arm, lie horizontal for 5 min, and then record BP;
- While still wearing the cuff, stand immediately and record the BP;
- Measure twice a day (on awakening and before arising and at night before retiring);
- Only perform measurements in the presence of a carer and sit or lie on the bed immediately if a risk of falling is perceived;
- After each reading, record the systolic and diastolic pressures on the provided chart. To avoid bias, PwPs were not informed about the meaning of the BP parameters they recorded.

Both the partner and PwP were asked to attend the training session and nominate which of them would be responsible for the recordings. The carer took responsibility approximately 50% of the time, particularly when cognition of the PwP was affected. Subjects were shown how to perform the recordings and how to record the result on the chart provided. They were requested to perform BP recordings until competent.

All control subjects and 84% of PwPs recorded BP on 5 consecutive days, with the remaining 16% making recordings on 4 of the 5 days. The difference between standing and lying systolic BP (ΔBP) was calculated: a positive number indicated standing BP > lying BP. While ΔBP described the difference between a single pair of measurements, there were 10 measurement pairs (ΔBP) made at home over 5 days. These were described by the median, 75th percentile (the 3rd highest of 10 ΔBP), and the maximum of the 10 measurements (notated as ΔBP_{med}, ΔBP_{75th}, and ΔBP_{MAX}, respectively). A ΔBP equal to or greater than 20 mmHg was considered "high". Systolic readings were defined as hypertensive if they were equal to or greater than 145 mmHg.

2.4. Statistics

As most distributions did not pass the D'Agostino and Pearson normality test and populations were small, the null hypothesis for the two distributions was tested using the Mann–Whitney test or the Wilcoxon matched pairs signed-rank test when the data were-paired. Categorical comparisons were assessed using the chi-squared test (or Fisher's exact test if the samples were small). Cohen's kappa statistic was used to measure concordance between existing measures of orthostatic hypotension and those from 5 days of recording at home. Statistical significance was set at 0.05.

3. Results

3.1. Characteristics of Morning and Evening Systolic BP Readings

The median lying systolic and diastolic BP of the PwPs and controls are shown in Table 1. However, aggregating the readings obscures detail revealed by examining all systolic BP readings (432 from the PwPs and 160 from the controls) (Figure 1A). As 85% of participants contributed ten readings and the minimum from any subject was eight readings, all participants provided similar amounts of data, and examining every recording (as in Figure 1A) was not biased by one individual's data. The median systolic BP reading was hypertensive (≥145 mmHg) in 32% of the PwPs and 25% of the controls.

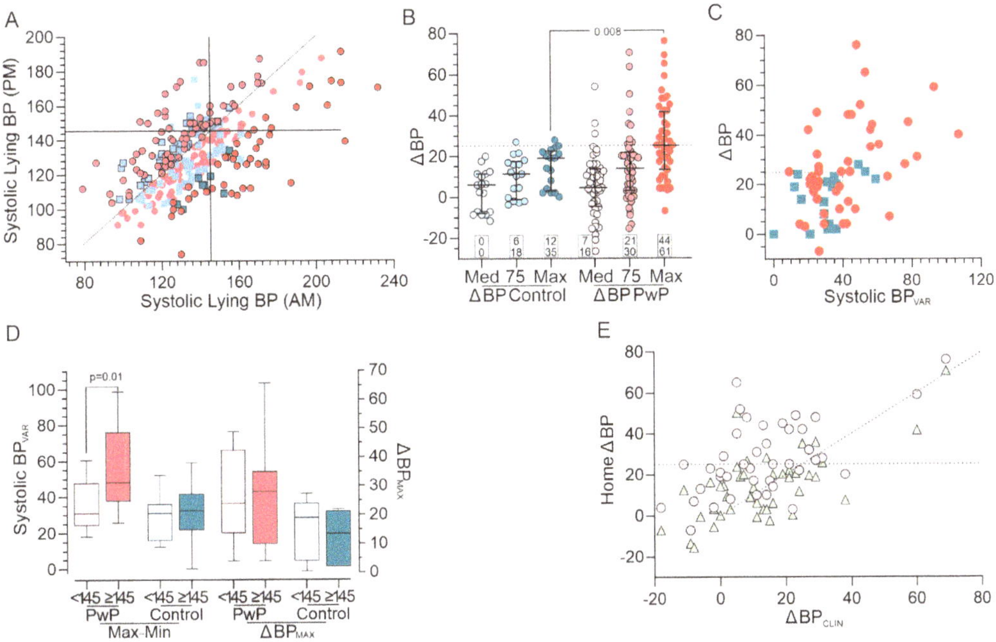

Figure 1. (**A**) A plot of the morning (AM, x-axis) and evening (PM, y-axis) lying systolic BP. Vertical and horizontal black lines indicate a reading of 145 mmHg and the dotted line indicates when morning and evening readings are the same. Pink circles show measurements from the PwP, with circles with a black border indicating cases where the evening reading was greater than the morning reading. Red circles indicate the PwP whose morning systolic measurement was ≥20 mmHg higher than the evening measurement. Grey squares show measurements from the controls, and those with a black border are cases whose evening reading was greater than the morning reading. Teal squares indicating cases where the evening reading was greater than the morning reading; (**B**) scatter plots (error bars: median and IQR) of the ΔBP (y-axis) sorted according to each participant's ΔBP_{med}, ΔBP_{75th}, and ΔBP_{MAX} (each circle indicates an individual participant). The grey-shaded region represents a ΔBP of 20 mmHg, and the horizontal dotted line indicates ΔBP of 25 mmHg. At the base of each plot are two sets of numbers in a box: the lower number indicates the percentage of that category where the ΔBP ≥ 20 mmHg and the upper row indicates the percentage of that category where the ΔBP ≥ 25 mmHg. Only p-values < 0.05 (Mann–Whitney test) are shown; (**C**) a plot of ΔBP_{MAX} (y-axis) of the PwPs (red circles) and the controls (teal squares) against the difference between the Syst BP_{Var} (x-axis: maximum–minimum lying systolic BP). The grey shaded area represents the region where both the ΔBP_{MAX} < 20 and the Syst BP_{Var} < 40 mmHg (which is~ the 75th percentile of the controls; see (D)) are present; (**D**) box (median and interquartile range) and whiskers (10th and 90th percentile) representing the range of Syst BPVar (left y-axis: maximum–minimum systolic BP) and ΔBP_{MAX} (right y-axis) of the PwPs (pink and red boxes) and the controls (grey and teal). Only p-values < 0.05 (Mann–Whitney test) are shown; (**E**) a plot of ΔBP_{75th} (green triangles) and ΔBP_{MAX} (pink circles) on the y-axis against the ΔBP_{CLIN} (x-axis). Concordance between ΔBP_{CLIN} and the measurements at home are symbols in the lower left grey rectangle (no OH) and the upper right grey rectangle (OH). Symbols in the upper left quadrant show cases where the home measurement detected OH but the clinic measurement did not, whereas symbols in the lower right quadrant show cases where the clinic measurement found OH but the home measures did not.

Morning and evening lying systolic BP from the same day were examined as a pair, leading to the following observations:

- *Morning systolic lying pressures are higher than their evening pair* in both PwP pairs (67%) and control pairs (75%). The difference between morning and evening systolic pressures was significant for both the PwPs (median difference = 6 mmHg, $p < 0.0001$—Wilcoxon matched pairs signed-rank test) and the controls (median difference = 4 mmHg, $p < 0.01$—Wilcoxon matched pairs signed-rank test);
- *If the morning lying systolic BP was 20 mmHg higher than its evening pair, it was frequently hypertensive in PwPs (78%) but not controls (38%).* On the other hand, when the evening lying systolic reading was the highest of the pair, the morning systolic was below 145 mmHg (80% of the PwPs and 98% of the controls).

3.2. Orthostatic Effects on Systolic BP

Measurements of standing and lying BP in the morning and evening for 5 days at home were used to calculate the ΔBP_{med}, ΔBP_{75th}, and ΔBP_{MAX} as measurements for evidence of OH (Figure 1B). Two observations are apparent. First, the proportion of subjects with a high ΔBP (by any of the three measures) was greater in the PwPs than in the controls (15.6%, 33.3%, and 62.2% for the PwPs and 0%, 17.6%, and 35.3% for the controls, respectively). Because of the number of controls with an elevated ΔBP_{MAX}, the effect of a higher threshold (for example, 25 mmHg being the 90th percentile of the controls) was also examined. The horizontal dotted line in Figure 1B shows this number and the number of cases whose $\Delta BP_{MAX} \geq 25$ is shown as the top number in the small boxes at the base of each graph.

Second, the variability in readings from the PwPs was greater than the variability of the controls (also apparent in Figure 1A). This variability was examined further by calculating the difference between the maximum and minimum morning and evening lying systolic BP (Syst BP_{Var}), which was considerably greater in the PwPs (39.5 (IQR = 30.3)) than in the controls ($p = 0.02$: unpaired t-test with Welch's correction). Syst BP_{Var} was plotted against ΔBP_{MAX} (Figure 1C), showing a modest relationship between the two measures (with Cohen's $\kappa = 0.51$ (discussed further below). This suggests that Syst BP_{Var} might be a marker of autonomic dysregulation, so it was compared in subjects with and without hypertension (Figure 1D). The Syst BP_{Var} was significantly larger in the PwPs when the median morning lying systolic BP was hypertensive ($p = 0.01$, Mann–Whitney test); this was not apparent in the controls. The trend for a higher orthostatic drop in hypertensive PwPs was not significant (Figure 1D), although PwPs with a large ΔBP_{MAX} ($\geq$20) had a higher systolic BP (147 (IQR = 39)) than those whose ΔBP_{MAX} was low (125 (IQR = 34), $p < 0.07$ t-test).

The interrelatedness of ΔBP_{MAX}, Syst BP_{Var}, and systolic BP was further examined. Of the 28 (out of 44) PwPs with an elevated ΔBP_{MAX}, 17 had a high Syst BPVar and 10 had hypertension. Hypertension or a high Syst BP_{Var} without a high BP_{MAX} was uncommon (9%). This suggests that these three measures are largely coincident.

3.3. Measurement of ΔBP at Home Compared to the Clinic

Next, the single office-based measurement of ΔBP (ΔBP_{CLIN}) was compared with ΔBP_{75th} and ΔBP_{MAX} (Figure 1E). The BP_{CLIN} was equal to or above 20 mmHg in 28.9% of the PwPs, which is a little less than ΔBP_{75th} (33.3%). ΔBP_{75th} was also a little better correlated with ΔBP_{CLIN} (Pearson's $\rho = 0.66$ and Cohen's $\kappa = 0.48$) than ΔBP_{MAX} (Pearson's $\rho = 0.58$ and Cohen's $\kappa = 0.34$). However, ΔBP_{75th} gave more "false negatives" (cases in the bottom right quadrant in Figure 1E where ΔBP_{75th} failed to detect the OH observed in the clinic) than ΔBP_{MAX}, whose differences with ΔBP_{CLIN} were almost all "false positives" (cases in the upper left quadrant in Figure 1E where ΔBP_{MAX} detected OH which was not observed in the clinic). It seems more plausible that one of the ten measures (ΔBP_{MAX}) would detect intermittent OH more accurately than either one of the seven measures (ΔBP_{75th}) or a single random measure in the clinic. For this reason, ΔBP_{MAX} was compared with scores from various clinical scales.

3.4. Relationship between ΔBP_{MAX} and Scores from Clinical Scales

The relationships between responses to Q1.12 of the MDS-UPDRS (light headedness on standing) and ΔBP_{MAX} and ΔBP_{75} are shown in Table 2. There was a progressive (but not statistically significant) trend for ΔBP_{MAX} to increase with a higher score to Q1.12. It was significant that a little more than half of those who responded with a "0" to this question had an elevated ΔBP_{MAX} and 20% of those who responded with a "2" or "3" had $\Delta BP_{MAX} < 20$. A higher Q1.12 score tended to be associated with a higher ΔBP_{MAX} (Figure 2A), even though ΔBP_{MAX} weakly predicted any answer of "1" or more to this question (Cohen's $\kappa = 0.23$). As a higher total score on the NMS-Quest scales was also associated with a higher ΔBP_{MAX} (Figure 2A), the relationship between ΔBP_{MAX} and responses to NMS questions specific to autonomic dysfunction (5–9, 19, 20, and 28) was examined (Figure 2A). An MDS-UPDRS I (total) score of 10 or more was associated with a higher ΔBP_{MAX} ($p = 0.026$, Mann–Whitney; see Figure 2B). There was a statistically insignificant trend for a lower MoCA and higher UPDRS III score with high ΔBP_{MAX}. No relationship was found between ΔBP_{MAX} and the PDQ39 or other MDS-UPDRS scores.

Table 2. The relationships between responses to Q1.12 of the MDS-UPDRS and ΔBP_{MAX} and ΔBP_{75}.

MDS-UPDRS Q1.12 Response	0	1	2	3
Number (%)	25 (57%)	9 (21%)	7 (16%)	3 (7%)
Median ΔBP_{MAX}	21 (20.5)	29 (35)	35 (36)	42 (41)
Median ΔBP_{75th}	9 (20)	19 (25)	26 (39)	19 (26)

0: Normal: No dizzy or foggy feelings. 1: Slight: Dizzy or foggy feelings occur. However, they do not cause me trouble doing things. 2: Mild: Dizzy or foggy feelings cause me to hold on to something, but I do not need to sit or lie back down. 3: Moderate: Dizzy or foggy feelings cause me to sit or lie down to avoid fainting or falling.

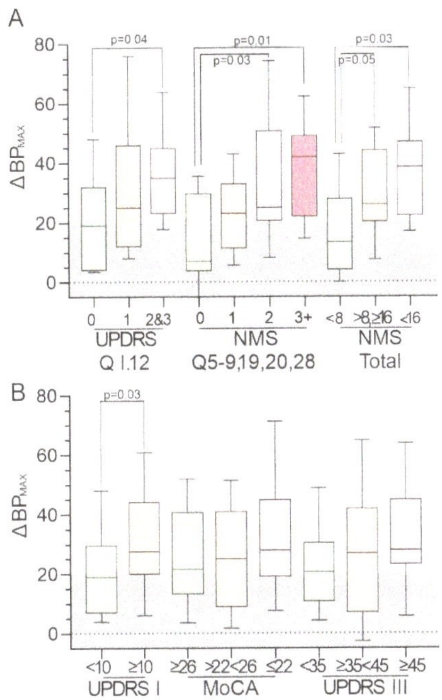

Figure 2. (**A**,**B**) are boxes (median and interquartile range) and whiskers (10th and 90th percentile), representing the ΔBP_{MAX} (y-axis) of various clinical scales in PwPs (abbreviations provided in the Methods). The grey shaded area indicates the region where the $\Delta BP_{MAX} < 20$ mmHg is found. Only p-values < 0.05 are shown from ordinary one-way ANOVA (in cases of three sets of data) or the Mann–Whitney test (in the case of two sets of data).

4. Discussion

The aim of this study was to assess whether more frequent measures of BP would provide a better indication of the presence of OH in PD. It was not intended to be a study of the incidence of OH in PD. PwPs with six or more years of disease duration were recruited to ensure that the study cohort included PwPs with and without OH. In the context of the aim of the study, its outcome can be assessed from a narrow perspective of whether an orthostatic drop in systolic BP was present and from a broader perspective of whether the dysregulation of systolic blood pressure control was present (expressed as morning hypertension and systolic BP variability (Syst BP_{Var}) and OH).

From a narrow perspective, multiple measurements were more likely to provide at least one measurement $\geq$ 20 mmHg (n = 28) than a single clinic measurement (n = 13). There was a significantly higher chance of having at least one elevated ΔBP in the PwPs than in the controls (p = 0.0026, Fisher's exact). Much of the thinking about OH is influenced by findings in hypovolaemia and OH induced by antihypertensive agents, which also provided the origin of a "high ΔBP" being $\geq$20 mmHg. In that setting, OH is expected to be consistently present. In contrast, OH in PD is intermittent [37], possibly reflecting central dysregulation [1,36–38] and the coincidence of different stressors, such as enteric shunting following food, and pharmaceuticals such as levodopa, hot environments, exercise, or hypovolaemia. While it seems logical that more frequent measurement would detect OH, it is unclear whether the optimum number of measurements should be five in ten (ΔBP_{MED}), three in ten (ΔBP_{75th}), one in ten (ΔBP_{MAX}), or even one in twenty. The findings of Polverino et al. [37] provide some indication that around 10 measurements may be sufficient, although all participants in that study had established OH. In this study, only 27% were recognised as having OH, yet 66% had at least one elevated ΔBP. Certainly, the association between ΔBP_{MAX} and worsening UPDRS I and NMS-Quest scores in this study suggests that OH detected by ΔBP_{MAX} is meaningful. Other than the study of Polverino et al. [37], we are not aware of a similar attempt to use conventional BP measurements of lying and standing BPs at home to assess the presence and severity of OH. The Polverino et al. study [37] used sophisticated telemetry, which may lend greater certainty to compliance, but the technology did not obviate the need for the PwP to interrupt their day for the length of time that is required to carry out lying and standing BPs. Also, they measured eight subjects with known OH and thus cannot provide an indication of the value of at-home measurements capturing milder forms of OH.

From a broader perspective, 10 readings at home provided insights regarding morning hypertension and increased variation in systolic pressures. However, hypertension or a high Syst BP_{Var} without a high ΔBP_{MAX} was uncommon (9%) and so, while a high Syst BP_{Var} or hypertension in the presence of a high ΔBP_{MAX} gives support to the finding of cardiovascular dysregulation, either Syst BPVar or hypertension in isolation does not.

While this study gives overall support for the home measurement of ΔBP, it does produce outstanding questions. Ten home measurements were arbitrary, as was the choice to perform the measurements at the start and end of the day rather than after meals. However, previous reports suggest that PwPs may not be fully aware of the presence of OH and misinterpret other symptoms as OH [37]. And, while the response to Q12.1 broadly correlated with the median postural drop, it was possible for individual cases to be unaware of OH, whereas others over-reported it. Early morning was chosen to capture supine hypertension. Measuring at the start and end of the day was expected to provide good compliance, which was excellent in this study, whereas asking subjects to measure when symptomatic may lead to overlooked measurements or measurements in response to irrelevant symptoms. Most participants (or their carer) had little difficulty in correctly measuring lying and standing BP, although subjects with cognitive impairment would not have been able to participate without a supportive carer. In summary, 10 recordings at home were superior to a single office measure in identifying the presence of OH as well as dysregulation in terms of a high Syst BP_{Var}. Nevertheless, the ideal number of measurements and their best timing throughout the day require further study.

We have used the conventional threshold of 20 mmHg or higher for an elevated ΔBP. Our definition was limited to a measurement immediately after standing, not 3 min later or even 10–30 min after standing. There was concern that a more complex measurement paradigm invited poorer compliance. Justification for the protocol used here was provided by the relationship between BP_{MAX} and the clinical scales, but in particular, those questions from the NMS Quest indicated autonomic dysfunction (Figure 2A). A systolic BP $\geq$145 mmHg was used to define hypertension. This was chosen as closer to the threshold that triggers concern in the real-world management of hypertension when OH is present, even though lower pressures may invoke intervention in otherwise healthy individuals. As morning hypertension was used as a proxy for supine hypertension, 145 mmHg may be too rigorous for some [42] but not for other [43] authorities.

Would early therapeutic intervention result in a more sensitive measure indicator of the presence of OH? While the relationship between OH and cognitive decline has led to recommendations of early interventions [44], it is not clear whether OH is a surrogate for supine OH (reviewed in [43]), although findings here suggest they usually co-exist. Others raise the possibility that OH and cognitive decline are para-phenomena [20].

In this study, people with insulin-dependent diabetes were excluded to avoid other causes of autonomic dysfunction. Participants using diuretics were excluded because diuretics might exacerbate hypovolaemia, but the use of antihypertensives was not grounds for exclusion. We acknowledge this is inconsistent, especially as the use of antihypertensives was higher in the controls than in the PwPs, suggesting that antihypertensives might have been explicitly avoided in PwPs. As diabetes and the use of diuretics and hypertensives are nearly ubiquitous in this age group, a future study could accept their presence because it reflects the complexity facing the management of Parkinsonian subjects with autonomic dysregulation and other conditions. Further studies on the concurrent management of OH and these conditions are required, especially for morning hypertension.

Limitations of This Study

Many of the limitations of this study have been addressed above. These and other limitations are summarised here:

- Are 10 measures adequate or too few? Should measures at other times (e.g., postprandial) also be included?;
- What proportion of measures should be sufficient to identify OH: 50% (ΔBP_{MED}), 33% (ΔBP_{75th}), 10% (ΔBP_{MAX}), 5%, or even less?;
- This study did not use the more stringent criteria for OH and systolic hypertension recommended by some authorities;
- This study excluded insulin-dependent diabetes and users of diuretics but not users of antihypertensive agents. However, because of the loss of ability to regulate vasodilation in the various vascular beds, it is these cases that introduce complexity to the management of OH in PD. Thus, future studies could examine the trade-off in treating hypertension in the presence of OH, especially when multiple measurements, such as those proposed here, are used;
- Although participants were trained to use the sphygmomanometers, we cannot exclude the possibility that some recordings were the result of poor technique or inaccurate recording. Poor technique might over-report hypotension and could also under-represent large postural drops. It is notable that very few systolic BP measures were less than 100 mmHg (Figure 1A);
- The sample size was large enough to show that at least one elevated ΔBP in 10 measurements is more likely to be found in PD than the controls. Larger samples would be required to address the dot points outlined above.

5. Conclusions

Twice daily recordings of lying and standing BP over 5 days increase the likelihood of finding an elevated postural drop, which is consistent with OH. Moreover, it can show whether there is increased variation in systolic pressures and morning hypertension, which is consistent with the dysregulated control of BP. PwPs were able to perform and record BP measurements without complications, were compliant, and did not find it intrusive.

Further studies are required to establish whether 5 days of recording is sufficient and whether active intervention on finding dysregulated BP control results in better outcomes compared to waiting for symptomatic OH before intervening.

Author Contributions: Conceptualisation, M.H.; methodology, C.V.F., S.O. and M.H., formal analysis, C.V.F., S.O. and M.H., investigation, C.V.F. and S.O.; resources, M.H.; data curation, C.V.F. and S.O.; writing—original draft preparation, M.H.; writing—review and editing, C.V.F. and S.O.; funding acquisition, M.H. All authors have read and agreed to the published version of the manuscript.

Funding: This research was funded by Global Kinetics: https://www.pkgcare.com/ (accessed on 28 August 2003).

Institutional Review Board Statement: This study was approved and overseen by the St. Vincent's Hospital (Melbourne) Human Research and Ethics Committee (approval number LRR 320.21). The subjects provided written consent according to the Declaration of Helsinki, and this study was conducted according to the International Conference on Harmonisation: Good Clinical Practice Guidelines.

Informed Consent Statement: Written informed consent was obtained from all subjects involved in this study.

Data Availability Statement: The data are available upon reasonable request to the corresponding author.

Conflicts of Interest: Global Kinetics (GK) provided funding for this study. C.F. and S.O. have no financial interests in GK. M.H. has equity in GK and has been the CSO and acting CEO of GK. The funders had no role in the design of the study; in the collection, analyses, or interpretation of data; in the writing of the manuscript; or in the decision to publish the results.

References

1. Velseboer, D.C.; de Haan, R.J.; Wieling, W.; Goldstein, D.S.; de Bie, R.M. Prevalence of orthostatic hypotension in Parkinson's disease: A systematic review and meta-analysis. *Park. Relat. Disord.* **2011**, *17*, 724–729. [CrossRef] [PubMed]
2. Palma, J.A.; Kaufmann, H. Epidemiology, Diagnosis, and Management of Neurogenic Orthostatic Hypotension. *Mov. Disord. Clin. Pract.* **2017**, *4*, 298–308. [CrossRef] [PubMed]
3. Palma, J.A.; Gomez-Esteban, J.C.; Norcliffe-Kaufmann, L.; Martinez, J.; Tijero, B.; Berganzo, K.; Kaufmann, H. Orthostatic hypotension in Parkinson disease: How much you fall or how low you go? *Mov. Disord.* **2015**, *30*, 639–645. [CrossRef] [PubMed]
4. Tipton, P.W.; Cheshire, W.P. Mechanisms underlying unawareness of neurogenic orthostatic hypotension. *Clin. Auton. Res.* **2020**, *30*, 279–281. [CrossRef]
5. Wieling, W.; Kaufmann, H.; Claydon, V.E.; van Wijnen, V.K.; Harms, M.P.M.; Juraschek, S.P.; Thijs, R.D. Diagnosis and treatment of orthostatic hypotension. *Lancet Neurol.* **2022**, *21*, 735–746. [CrossRef]
6. Juraschek, S.P.; Daya, N.; Rawlings, A.M.; Appel, L.J.; Miller, E.R., 3rd; Windham, B.G.; Griswold, M.E.; Heiss, G.; Selvin, E. Association of History of Dizziness and Long-term Adverse Outcomes with Early vs Later Orthostatic Hypotension Assessment Times in Middle-aged Adults. *JAMA Intern. Med.* **2017**, *177*, 1316–1323. [CrossRef]
7. Finucane, C.; O'Connell, M.D.; Donoghue, O.; Richardson, K.; Savva, G.M.; Kenny, R.A. Impaired Orthostatic Blood Pressure Recovery Is Associated with Unexplained and Injurious Falls. *J. Am. Geriatr. Soc.* **2017**, *65*, 474–482. [CrossRef]
8. Shaw, B.H.; Claydon, V.E. The relationship between orthostatic hypotension and falling in older adults. *Clin. Auton. Res.* **2014**, *24*, 3–13. [CrossRef]
9. Juraschek, S.P.; Daya, N.; Appel, L.J.; Miller, E.R., 3rd; Windham, B.G.; Pompeii, L.; Griswold, M.E.; Kucharska-Newton, A.; Selvin, E. Orthostatic Hypotension in Middle-Age and Risk of Falls. *Am. J. Hypertens.* **2017**, *30*, 188–195. [CrossRef]
10. Tanaka, R.; Hattori, N. Abnormal circadian blood pressure regulation and cognitive impairment in alpha-synucleinopathies. *Hypertens. Res.* **2022**, *45*, 1908–1917. [CrossRef]
11. Frewen, J.; Savva, G.M.; Boyle, G.; Finucane, C.; Kenny, R.A. Cognitive performance in orthostatic hypotension: Findings from a nationally representative sample. *J. Am. Geriatr. Soc.* **2014**, *62*, 117–122. [CrossRef]
12. Frewen, J.; Finucane, C.; Savva, G.M.; Boyle, G.; Kenny, R.A. Orthostatic hypotension is associated with lower cognitive performance in adults aged 50 plus with supine hypertension. *J. Gerontol. A Biol. Sci. Med. Sci.* **2014**, *69*, 878–885. [CrossRef]
13. O'Hare, C.; Kenny, R.A.; Aizenstein, H.; Boudreau, R.; Newman, A.; Launer, L.; Satterfield, S.; Yaffe, K.; Rosano, C.; Health, A.B.C.S. Cognitive Status, Gray Matter Atrophy, and Lower Orthostatic Blood Pressure in Older Adults. *J. Alzheimers Dis.* **2017**, *57*, 1239–1250. [CrossRef]

14. Holm, H.; Nagga, K.; Nilsson, E.D.; Melander, O.; Minthon, L.; Bachus, E.; Fedorowski, A.; Magnusson, M. Longitudinal and postural changes of blood pressure predict dementia: The Malmo Preventive Project. *Eur. J. Epidemiol.* **2017**, *32*, 327–336. [CrossRef] [PubMed]
15. Hayakawa, T.; McGarrigle, C.A.; Coen, R.F.; Soraghan, C.J.; Foran, T.; Lawlor, B.A.; Kenny, R.A. Orthostatic Blood Pressure Behavior in People with Mild Cognitive Impairment Predicts Conversion to Dementia. *J. Am. Geriatr. Soc.* **2015**, *63*, 1868–1873. [CrossRef]
16. Kang, S.H.; Chung, S.J.; Lee, J.; Koh, S.B. Independent effect of neurogenic orthostatic hypotension on mild cognitive impairment in Parkinson's disease. *Clin. Auton. Res.* **2022**, *32*, 43–50. [CrossRef]
17. Zhou, Y.; Ke, S.J.; Qiu, X.P.; Liu, L.B. Prevalence, risk factors, and prognosis of orthostatic hypotension in diabetic patients: A systematic review and meta-analysis. *Medicine* **2017**, *96*, e8004. [CrossRef] [PubMed]
18. Lagro, J.; Schoon, Y.; Heerts, I.; Meel-van den Abeelen, A.S.; Schalk, B.; Wieling, W.; Olde Rikkert, M.G.; Claassen, J.A. Impaired systolic blood pressure recovery directly after standing predicts mortality in older falls clinic patients. *J. Gerontol. A Biol. Sci. Med. Sci.* **2014**, *69*, 471–478. [CrossRef]
19. Fedorowski, A.; Stavenow, L.; Hedblad, B.; Berglund, G.; Nilsson, P.M.; Melander, O. Orthostatic hypotension predicts all-cause mortality and coronary events in middle-aged individuals (The Malmo Preventive Project). *Eur. Heart J.* **2010**, *31*, 85–91. [CrossRef] [PubMed]
20. McDonald, C.; Newton, J.L.; Burn, D.J. Orthostatic hypotension and cognitive impairment in Parkinson's disease: Causation or association? *Mov. Disord.* **2016**, *31*, 937–946. [CrossRef]
21. Groothuis, J.T.; Esselink, R.A.; Seeger, J.P.; van Aalst, M.J.; Hopman, M.T.; Bloem, B.R. Lower vascular tone and larger plasma volume in Parkinson's disease with orthostatic hypotension. *J. Appl. Physiol.* **2011**, *111*, 443–448. [CrossRef] [PubMed]
22. Jacob, G.; Ertl, A.C.; Shannon, J.R.; Furlan, R.; Robertson, R.M.; Robertson, D. Effect of standing on neurohumoral responses and plasma volume in healthy subjects. *J. Appl. Physiol.* **1998**, *84*, 914–921. [CrossRef]
23. Sulzer, D.; Surmeier, D.J. Neuronal vulnerability, pathogenesis, and Parkinson's disease. *Mov. Disord.* **2013**, *28*, 715–724. [CrossRef]
24. Papapetropoulos, S.; Mash, D.C. Insular pathology in Parkinson's disease patients with orthostatic hypotension. *Park. Relat. Disord.* **2007**, *13*, 308–311. [CrossRef] [PubMed]
25. Pavelic, A.; Krbot Skoric, M.; Crnosija, L.; Habek, M. Postprandial hypotension in neurological disorders: Systematic review and meta-analysis. *Clin. Auton. Res.* **2017**, *27*, 263–271. [CrossRef]
26. Claydon, V.E.; Steeves, J.D.; Krassioukov, A. Orthostatic hypotension following spinal cord injury: Understanding clinical pathophysiology. *Spinal Cord.* **2006**, *44*, 341–351. [CrossRef] [PubMed]
27. Goldstein, D.S.; Eldadah, B.A.; Holmes, C.; Pechnik, S.; Moak, J.; Saleem, A.; Sharabi, Y. Neurocirculatory abnormalities in Parkinson disease with orthostatic hypotension: Independence from levodopa treatment. *Hypertension* **2005**, *46*, 1333–1339. [CrossRef]
28. Goldstein, D.S.; Sharabi, Y. Baroreflex-sympathoneural dysfunction characterizes at-risk individuals with preclinical central Lewy body diseases. *Clin. Auton. Res.* **2023**, *33*, 41–49. [CrossRef]
29. Kramer, H.H.; Lautenschlager, G.; de Azevedo, M.; Doppler, K.; Schanzer, A.; Best, C.; Oertel, W.H.; Reuter, I.; Sommer, C.; Birklein, F. Reduced central sympathetic activity in Parkinson's disease. *Brain Behav.* **2019**, *9*, e01463. [CrossRef]
30. Barbic, F.; Perego, F.; Canesi, M.; Gianni, M.; Biagiotti, S.; Costantino, G.; Pezzoli, G.; Porta, A.; Malliani, A.; Furlan, R. Early abnormalities of vascular and cardiac autonomic control in Parkinson's disease without orthostatic hypotension. *Hypertension* **2007**, *49*, 120–126. [CrossRef]
31. Senard, J.M.; Valet, P.; Durrieu, G.; Berlan, M.; Tran, M.A.; Montastruc, J.L.; Rascol, A.; Montastruc, P. Adrenergic supersensitivity in parkinsonians with orthostatic hypotension. *Eur. J. Clin. Invest.* **1990**, *20*, 613–619. [CrossRef] [PubMed]
32. Sharabi, Y.; Goldstein, D.S. Mechanisms of orthostatic hypotension and supine hypertension in Parkinson disease. *J. Neurol. Sci.* **2011**, *310*, 123–128. [CrossRef]
33. Cuenca-Bermejo, L.; Almela, P.; Navarro-Zaragoza, J.; Fernandez Villalba, E.; Gonzalez-Cuello, A.M.; Laorden, M.L.; Herrero, M.T. Cardiac Changes in Parkinson's Disease: Lessons from Clinical and Experimental Evidence. *Int. J. Mol. Sci.* **2021**, *22*, 13488. [CrossRef]
34. Zhu, S.; Li, H.; Xu, X.; Luo, Y.; Deng, B.; Guo, X.; Guo, Y.; Yang, W.; Wei, X.; Wang, Q. The Pathogenesis and Treatment of Cardiovascular Autonomic Dysfunction in Parkinson's Disease: What We Know and Where to Go. *Aging Dis.* **2021**, *12*, 1675–1692. [CrossRef]
35. Mehagnoul-Schipper, D.J.; Boerman, R.H.; Hoefnagels, W.H.; Jansen, R.W. Effect of levodopa on orthostatic and postprandial hypotension in elderly Parkinsonian patients. *J. Gerontol. A Biol. Sci. Med. Sci.* **2001**, *56*, M749–M755. [CrossRef]
36. Xing, Y.; Li, Q.; Xu, E.; Zeng, J.; Li, Q.; Mei, S.; Hua, Y. Impaired Cerebral Autoregulation in Parkinson's Disease: An Orthostatic Hypotension Analysis. *Front. Neurol.* **2022**, *13*, 811698. [CrossRef] [PubMed]
37. Polverino, P.; Ajcevic, M.; Catalan, M.; Bertolotti, C.; Furlanis, G.; Marsich, A.; Buoite Stella, A.; Accardo, A.; Manganotti, P. Comprehensive telemedicine solution for remote monitoring of Parkinson's disease patients with orthostatic hypotension during COVID-19 pandemic. *Neurol. Sci.* **2022**, *43*, 3479–3487. [CrossRef] [PubMed]
38. Vallelonga, F.; Di Stefano, C.; Merola, A.; Romagnolo, A.; Sobrero, G.; Milazzo, V.; Burrello, A.; Burrello, J.; Zibetti, M.; Veglio, F.; et al. Blood pressure circadian rhythm alterations in alpha-synucleinopathies. *J. Neurol.* **2019**, *266*, 1141–1152. [CrossRef] [PubMed]

39. Lodhi, H.A.; Peri-Okonny, P.A.; Schesing, K.; Phelps, K.; Ngo, C.; Evans, H.; Arbique, D.; Price, A.L.; Vernino, S.; Phillips, L.; et al. Usefulness of Blood Pressure Variability Indices Derived From 24-Hour Ambulatory Blood Pressure Monitoring in Detecting Autonomic Failure. *J. Am. Heart Assoc.* **2019**, *8*, e010161. [CrossRef]
40. Freeman, R.; Illigens, B.M.W.; Lapusca, R.; Campagnolo, M.; Abuzinadah, A.R.; Bonyhay, I.; Sinn, D.I.; Miglis, M.; White, J.; Gibbons, C.H. Symptom Recognition Is Impaired in Patients With Orthostatic Hypotension. *Hypertension* **2020**, *75*, 1325–1332. [CrossRef]
41. Chaudhuri, K.R.; Martinez-Martin, P.; Schapira, A.H.; Stocchi, F.; Sethi, K.; Odin, P.; Brown, R.G.; Koller, W.; Barone, P.; MacPhee, G.; et al. International multicenter pilot study of the first comprehensive self-completed nonmotor symptoms questionnaire for Parkinson's disease: The NMSQuest study. *Mov. Disord.* **2006**, *21*, 916–923. [CrossRef] [PubMed]
42. Fanciulli, A.; Jordan, J.; Biaggioni, I.; Calandra-Buonaura, G.; Cheshire, W.P.; Cortelli, P.; Eschlboeck, S.; Grassi, G.; Hilz, M.J.; Kaufmann, H.; et al. Consensus statement on the definition of neurogenic supine hypertension in cardiovascular autonomic failure by the American Autonomic Society (AAS) and the European Federation of Autonomic Societies (EFAS): Endorsed by the European Academy of Neurology (EAN) and the European Society of Hypertension (ESH). *Clin. Auton. Res.* **2018**, *28*, 355–362. [CrossRef] [PubMed]
43. Kaufmann, H.; Palma, J.A. White Matter Hyperintensities in the Synucleinopathies: Orthostatic Hypotension, Supine Hypertension, or Both? *Mov. Disord. Clin. Pract.* **2020**, *7*, 595–598. [CrossRef] [PubMed]
44. Merola, A.; Romagnolo, A.; Rosso, M.; Lopez-Castellanos, J.R.; Wissel, B.D.; Larkin, S.; Bernardini, A.; Zibetti, M.; Maule, S.; Lopiano, L.; et al. Orthostatic hypotension in Parkinson's disease: Does it matter if asymptomatic? *Park. Relat. Disord.* **2016**, *33*, 65–71. [CrossRef]

MDPI AG
Grosspeteranlage 5
4052 Basel
Switzerland
Tel.: +41 61 683 77 34

Journal of Personalized Medicine Editorial Office
E-mail: jpm@mdpi.com
www.mdpi.com/journal/jpm

www.ingramcontent.com/pod-product-compliance
Lightning Source LLC
Chambersburg PA
CBHW041954260926
51850CB00019B/2731
9783725852932